ISBN 978-0-331-82623-4
PIBN 10260070

1 MONTH OF
FREE
READING

at

www.ForgottenBooks.com

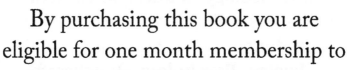

By purchasing this book you are eligible for one month membership to ForgottenBooks.com, giving you unlimited access to our entire collection of over 1,000,000 titles via our web site and mobile apps.

To claim your free month visit:

www.forgottenbooks.com/free260070

English
Français
Deutsche
Italiano
Español
Português

www.forgottenbooks.com

Mythology Photography **Fiction**
Fishing Christianity **Art** Cooking
Essays Buddhism Freemasonry
Medicine **Biology** Music **Ancient
Egypt** Evolution Carpentry Physics
Dance Geology **Mathematics** Fitness
Shakespeare **Folklore** Yoga Marketing
Confidence Immortality Biographies
Poetry **Psychology** Witchcraft
Electronics Chemistry History **Law**
Accounting **Philosophy** Anthropology
Alchemy Drama Quantum Mechanics
Atheism Sexual Health **Ancient History**
Entrepreneurship Languages Sport
Paleontology Needlework Islam
Metaphysics Investment Archaeology
Parenting Statistics Criminology
Motivational

SUMMARY

1892.

S. BAYL

NOTES IN THE "AMERICAN NATURALIST"

SUMMARY OF PROGRESS

IN

Mineralogy and Petrography

IN

1892.

BY

W. S. BAYLEY.

FROM MONTHLY NOTES IN THE "AMERICAN NATURALIST."

PRICE 50 CENTS.

WATERVILLE, ME.:
GEOLOGICAL DEPARTMENT, COLBY UNIVERSITY,
1893.

ε H.

135582

MINERALOGY AND PETROGRAPHY.[1]

Petrographical News.—The eruptive rocks of Velay, Haute Loire, France, in the order of their age are basalts, trachytes and trachytic phonolites, augite andesites, porphytic basalts, nepheline phonolites and nepheline basalts. Termier,[2] who describes them, gives but a few brief notes on each type. The younger phonolites form the larger part of the hill. They contain aegerine in light-green porphyritic crystals, and in microlites. At the south-east of St. Pierre-Eynac are tertiary clay slates cut by dykes of phonolite, whose tiny veins penetrate metamorphosed phases of the clastics, and are thus consequently regarded as the agents producing the alteration. The rocks representing the first stage in the alteration consist of granitic debris, in which secondary opal has been deposited around the feldspar and quartz fragments. In some instances, in addition to the opal there have been formed also secondary quartz and calcite, the former as a fibrous rim around the grains. In more intensely changed phases, the slate is traversed by veins of phonolite, whose contact with the sedimentary rock is not visible, since on both sides of it the material of the phonolite has thoroughly impregnated the slate. On the other hand the phonolite of the veins contains sphene, but no augite, while the normal rock contains an abundance of aegerine, but no sphene. In the final stage all the quartz of the slate has disappeared, and the rock is comprised principally of opal, serpentine and clay (halloysite?), with pleonaste, colorless augite and hornblende as new products. The alteration is thus a silicification. In other, more rare cases, it is a feldspathization.—Hutchings[3] has recently studied the material of which slates are formed, having examined for this purpose, clays and micaceous sand-

[1] Edited by Dr. W. S. Bayley, Colby University, Waterville, Me.

[2] Bull. d. Serv. d. l. carte Geol. d. Fr.   No. 13, 1890.

[3] Geological Magazine VII, 1890, pp. 264 and 316, and Ib. 1891, p. 164.

stones from near Seaton, England, north of Newcastle-on-Tyne. The harder shale layers are composed mainly of mica, quartz, feldspar, zircon and other accessory minerals, as garnet, rutile, anatase, tourmaline, sphene and barite. The biotite is the first of these to undergo change under the influence of weathering processes. In the case studied it has not changed to chlorite, but has become bleached and has yielded epidote. The quartz and feldspar are uniformly distributed throughout the mass, while the mica usually lies out its flat surfaces in the bedding planes. In addition to the mineral grains already mentioned there is present a sort of groundmass or paste, made up of indistinctly granular matter, with microlites of various kinds and a large amount of a fine micaceous substance, besides large plates of secondary mica. In a fine grained portion of the deposit, the paste is quite abundant and in it are numbers of minute rutile needles, flakes of ilmenite, some small perfect crystals of tourmaline and a considerable quantity of the secondary mica (muscovite), which differs from the clastic mica in that the edges of the grains extend out between the surrounding minerals, and the plates are full of minute rutile needles. In the very finest grained, smoothest clay bands of the region, the paste forms the largest part of the mass, while the well marked clastic grains are few in number, the biotite having disappeared entirely. Kaolin was not certainly recognized in even the thinnest sections of fire-clay, the fine grained granular substance of which this clay principally consists, being mainly the paste described above. The abundance of rutile, that is so noticeable a feature of the clays examined, is supposed to have originated upon the decomposition of the biotites, and the muscovite (or sericite), by the alteration of the paste. This mineral gradually increases in quantity, and then under the influence of pressure is so orientated that a micaceous slate results. The absence of biotite from most slates is thought to be due to the easy decomposability of the substance; and its presence in the sediments from which some slates were formed is thought to be indicated by the large percentage of rutile and epidote in the latter. In his second contribution to the subject the author describes the results of a separation of the components of a fine clay by fractional levigation, and an examination of the separated portions. He concludes from his study that nearly all the muscovite of slates, and all of the rutile bearing variety, is a secondary product, subsequent in origin to the deposition of the material from which the slates were formed.—The iron ores of Sao Paulo, Brazil, are found in two principal districts, the Jacupiranga and the Ipamena. In the first locality the ore, with a violet titaniferous pyroxene, forms a schistose rock, in which perofskite, apatite and a zeolitized silicate are accessory constituents. As the amount of the ore increases, that of the other constituents decreases, until in some cases an almost pure magnetite results. By weathering the augite gives rise to mica in abundance. In some cases the pyroxene rock is found associated with layers in which magnetite and nepheline are the principal components. This Derby[1] believes to be genetically connected with the ore-bearing rocks, which he calls jacupirangites, and regards as eruptive. Much of the ore of the Ipamena district originally occurred with acmite and apatite as segregations in acmite syenite, which is now highly decomposed, so that the segregations are scattered like boulders over the ground. Fouque[2] has redetermined the minerals of the Santorin rocks and has discovered that his former determinations of some of them were erroneous. In the pumice of Acroti, Isle of Thera, composed of glass fibres with opal material between, are little transparent crystals whose nature has heretofore been in doubt. A new examination proves them to be alunite. A quantity—separated and analyzed—gave: $SO_3=38\%$, $Al_2O_3=37.3\%$; $H_2O=13.3$; alkalies$=11.4\%$. The blocks enclosed in the lava of 1886, formerly supposed to consist of wollastinate, *fassaite* and melanite, are now known to contain in addition to these substances anhydrite. The interiors of the nodules are composed almost exclusively of anhydrite, with a little augite and other constituents of the enclosing rock, in which are wollastonite and melanite. —Four small boulders of nephrite from British Columbia have been examined by Harrington.[3] Three were found near Lytton on the Fraser River, and the fourth in the upper part of the Lewes River, near the Alaskan boundary line. The composition of each is as follows:

	SiO$_2$	Al$_2$O$_3$	FeO	MnO	CaO	MgO	Loss	Sp. Gr.
1.	55.32	2.42	5.35	.52	14.00	20.16	2.16	3.0278
2.	56.98	.18	4.59	.17	12.99	22.38	2.64	3.003
3.	56.54	.40	3.61	.16	13.64	22.77	2.92	3.01
4.	56.96	.51	3.81	.53	13.29	22.41	2.91	3.007

No. 3 contains pale hornblende crystals extinguishing at 8°-18°. —Derby[4] has discovered that xenotime is an almost universal constituent of muscovite granites, and presumably of other acid potash rocks. Residues obtained by washing the powder of such rocks in a gold washer's pan nearly always yields xenotime and monazite. Eighty-six and two-thirds percentage of the undoubted muscovite granites from

[1] Amer. Jour. Sci., Apr., 1891, p. 311.
[2] Bull. Soc. Franc. d' Min., 1890, XIII, p. 245.
[3] Trans. Roy. Soc. Can. 1890, p. 61.
[4] Amer. Jour. Sci., Apr., 1891, p. 308.

Brazil, examined by Derby, disclosed the presence of this accessory, which is thought by him to be as constant a constituent of these rocks as any mineral save zircon.   Experiments made with granites from the United States seem to indicate the value of the pan as a petrographical instrument in the study of the rare and heavy components of rocks. —The rare blue hornblende riebeckite is reported by Cole[1] in three pebbles found in the drift of North England and of Wales, in addition to its occurrences in the microgranite of Mynydd Mawr, where it was discovered by Harker and Bonney a few years ago.

Mineralogical News.—General.—Since many of the supposed paramorphs have been proven within recent years to be due not to the molecular rearrangement of material already existing, but rather to the solution of some original substance and its replacement by a new deposition, Bauer[2] has re-examined the pseudomorphs of *rutile* after *brookite* from Magnet Cove, Ark., to determine whether or not the substance is a true paramorph.   After studying many thin sections of the brookite, rutile, and intergrowths of the two, he concludes that the latter are true paramorphs, the rutile originating in a molecular re-arrangement of the TiO_2.   The rutile begins to form on the exterior of the brookite crystals, or along cracks in them, as needles penetrating the brookite substance.   The rutile pseudomorphs after anatase from Brazil, the Urals and other[3] localities, are also declared to be true paramorphs.   The same author has also re-examined the *Michel-levyite* of Lacroix, which Dana declared to be barite, and finds that Dana's statement is correct. The axial angle is large, but it cannot be measured, as the acute bisectrix does not enter the field of view.   The mineral differs from ordinary barite only in the possession of a very perfect cleavage in the direction of one prismatic face.   On the base it shows twinning striations resembling those of plagioclase.   The twinning plane is the prismatic face parallel to which is the most perfect cleavage.   The structural peculiarities of the Perkin's Mill barite are all due to this abnormal cleavage, which in turn is dependent upon the twinning, which is new to the mineral, and is probably the result of pressure.   Measurements of druse crystals of the same substance yield the forms characteristic of barite and the new plane $\frac{1}{12}$ P$\check{\infty}$ with $a: b: c: = .8152 : 1 : 1.3136$.—The mineral associated with calcite and phillipsite as druses on the leucite,

[1] Miner. Magazine, IX, 1891, p. 222.

[2] Neues. Jahrb. f. Min., etc., 1891, I, p. 217.

[3] Ib. p. 250.

tephrite of Eulenberg in Bohemia, pronounced by Zepharovich[1] to be orthoclase, has been carefully investigated by Gränzer,[2] who thinks it more likely to be a *zeolite*. Its crystallization is probably triclinic, though by the parallel growth of many individuals there is built up a form closely resembling that of orthoclase.—The minerals characterizing the hollow spherulites of the rhyolite of Glade Creek,[3] Wyoming, and of Obsidian Cliff, in the Yellowstone National Park, like those found in other lithophysæ, are thought to be the results of acqueo-igneous fusion upon the material of the acid lava. The most abundant mineral thus formed is *quartz*, whose crystals are either attached to the walls of the cavities, thus exposing only one termination, or are interlaced forming a network built up of crystals occasionally doubly-terminated. Both the rare $+$ ⅓ R and $-$ ⅓ R are well developed, and also the equally rare forms ±⅓ P⅓. The next most noticeable mineral is *fayalite*, whose habit has already[4] been described. In some of the more irregular cavities at Glade Creek are accumulations of very small *sanidine* crystals, *hornblende* and *biotite*, of which the latter is never found associated with fayalite.—The *rhodizite* from the Urals, which has been declared to be regular with ∞ O and $\frac{O}{2}$, is pyroelectric. The examination of it, extinction and its interference colors shows it to be pseudosymmetrical, it being in reality monoclinic[5] with $a: b: c=.707 : 1\ 1. \beta=90°$. The dodecahedron becomes OP,—P, $+$ P and ∞ P∞˘ and the tetrahedron P∞̄ and ∞ P 2'. An interesting series of experiments made by the same mineralogist on *jeremejewite* lately described by Websky[6] shows it to consist of an interior hexagonal kernel, surrounded by two zones with some orthorhombic properties, and an external one, with the characteristics of the kernel. The density of the material in each zone is the same, and its reaction under pressure and temperature is similarly slight. The kernel and the peripheral zones are uniaxial and negative, while the other two zones are biaxial, the inner one possessing the larger optical angle. The explanation of the phenomena offered by the author is to the effect that in the first stage of the mineral's growth it separated as an orthorhombic substance on the walls of the cavity. Upon this were deposited zones two and three, after which the cavity was filled by what is now the kernel. The optical anomalies of *phacolite*

[1] Sit. d. Kais. Ak. d. Wiss., 1885, X, p. 601.

[2] Min. u. Petrog. Mitth., XI, 1890, p. 277.

[3] Iddings & Penfield: Amer. Jour. Sci., July, 1891, p. 39.

[4] Ib. XXX, July, 1885, p. 59, and Ib. p. 271.

[5] Klein. Neues. Jahrb. f. Min., etc., 1891, I, p. 77.

[6] Ib. 1884, I, p. 1.

and *chabazite* are ascribed to the same causes as those assigned to the anomalies of analcite.—Some doubt having been cast upon the correctness of Baumhauer's conclusion that *nepheline* is trapezohedrally hemihedral, the mineral from three bombs of Vesuvius has again been examined. The figures produced on the ∞ P faces of crystals, upon etching with HCl and HF, are unsymmetrical; consequently their crystallization is either pyramidal or trapezohedral hemidedral, and the forms are hemimorphic with respect to the vertical axis.[1]—Saltman records the following as the composition of a *melanite* from Oberrothwell in the Kaiserstuhl:

SiO_2	TiO_2	ZrO_2	Al_2O_3	Fe_2O_3	Mn_2O_3	FeO	CaO	MgO	Na_2O	K_2O	Loss
30.48	11.01	1.28	3.13	15.21	.28	3.84	30.19	2.28	1.65		.19

It is interesting for the large percentage of titanium shown by it, and for the considerable quantity of zirconium, which has heretofore never been found in any member of the garnet group.—After examining critically more than fifty analyses of *vesuvianite*, Kenngott[2] concludes that the composition of the mineral must be represented by a formula of two parts, like that of apatite. The silicate portion may be represented by $4 (2 RO. SiO_2) + 2 R_2O_3, 3 SiO_2 [=4 R''_2 SiO_4 + R'''_4 (SiO_4)_3]$. The composition of the non silicated portion is not yet known, but it probably contains the hydroxyl group, sodium, potassium, and sometimes flourine, in varying proportions.—*Almogen*[4] crystals from the Pic-de-Teyde, Teneriff are tabular in habit. They are negative and crystallize monoclinically with $a: c=1 : .825.$ $\beta=97°34'$.—Rose colored dodecahedral garnets from Xalostic, Mex., have been analyzed by De Landero.[5] Their density is 3.516 and hardness 7.5. Their composition corresponding to $(Ca\ Mg)_3 (Al\ Fe)_2 (SiO_4)_3$, is:

SiO_2	Al_2O_3	Fe_2O_3	CaO	MgO	MnOBaO	Res.
40.64	21.48	1.57	35.38	.75	tr.	.17

—Some good sections of *pericline* from the Pfitschthal, Tyrole, have been very carefully studied by Münsig.[6] Their optical properties indicate that the substance is not a pure albite, but that it is an intergrowth of oligoclase (ab-an) with albite. The former comprehends the larger part of the pericline crystals, the latter appearing in it as small irregular flecks. Both feldspars are twinned according to the pericline law, with the albite apparently occupying pores in the oligoclase.—Des

[1] Baumhauer: Zeits. f. Kryst. XVIII, p. 611.
[2] Ib. p. 628.
[3] Neues. Jahrb. f. Min., etc., 1891, I, p. 200.
[4] Becke : Min. u. Petrog. Mitth., XII, p. 45.
[5] Amer. Jour. Sci., 141, 1891, p. 321.
[6] Neues. Jahrb. f. Min., etc., 1891, II, p. 1.

Cloizeaux[1] notes the similarity in habit between *chalcopyrite* crystals from Cuba and those of the French Creek Mines in Chester Co.,[2] Pa. A comparison of recent analyses of *violan* and *anthocroite* leads Igelström[3] to the conclusion that the two are identical.—In a recent brochure of the American Geological Society, Kunz[4] announces the discovery of small *diamonds* in the alluvial sands of Plum Creek, Pearce Co., Wis., and the occurrence of fire opal in a vesicular basalt at Whelan, Washington.—Sandberger[5] has found pseudomorphs of markasite after pyrargyrite at Chanarcillo, Chile.

Miscellaneous.—Syntheses.—Lorenz[6] has produced crystallized zinc sulphide by the sublimation of the amorphous salt in an atmosphere of ammonium chloride. The action is explained as taking place in two stages—first, the formation of zinc chloride and its sublimation, and second, the action of sulphuretted hydrogen upon this salt. By the action of dry H_2S on the respective metals crystalized *troilite, millerite, wurtzite* and *greenockite* were formed. The first is in little opaque tabular crystals, that are at first silver white and afterwards bronzy-yellow in color. According to Prof. Groth, they are probably hemimorphic. In addition to the greenockite there were produced in the same operation other crystals that are seemingly cadmium sulphide.—Though the synthetical production of augite is not a difficult problem, that of hornblende has heretofore resisted the best efforts of mineralogists to effect it. Chrustschoff[7] has however lately succeeded in obtaining the mineral by heating in a glass tube, from which the air had been extracted, a mixture of dialysed colloidal silicic acid containing 3% of SiO_2, and dialysed solutions of Al_2O_3, Fe (OH)$_3$ and Fe (OH)$_2$, with lime water, freshly prepared Mg (OH)$_2$ suspended in water, and a few drops of sodium and potassium hydroxides. Upon heating these together for about three months at 550° the mixture became of a dirty-brownish-green color, when it was found to contain tiny hard grains of hornblende, analcite quartz, feldspar and diopside. The hornblende crystals were bounded by $\infty P\infty$, $P\infty$ and ∞P. Their extinction $c \wedge C = 17°50'$. Their

[1]Bull. Soc. Franc. d. Min., XIII, p. 335.
[2]Cf. AMERICAN NATURALIST, 1889, p. 528.
[3]Neues. Jahrb. f. Min., etc., 1890, II, p. 271.
[4]Bull. Geol. Soc. Amer., Vol. 2, p. 638.
[5]Neues. Jahrb. f. Min., etc., 1891, I, p. 199.
[6]Ber. d. Deutsch. Chem. Ges. No. 9, 1891, p. 1501.
[7]Neues. Jahrb. f. Min., etc., 1891, II, p. 86.

double refraction was negative and pleochroism strong.   2 V=82°, and composition :

SiO_2	Al_2O_3	Fe_2O_3	FeO	MgO	CaO	Na_2O	K_2O	Loss
42.35	8.11	7.91	10.11	14.33	13.21	2.18	1.87	.91

—Otto and Kloos[1] find perfect crystals of periclase on a muffel in which magnesium oxychloride has been heated.

General.—The solubility of quite a number of minerals in pure water and in dilute salts has been carefully investigated by Doelter.[2]   The sulphides, sulpho-salts, oxides and silicates examined are slightly soluble in water, with the addition of sodium sulphide the solubility of the first two groups is increased while that of the oxides is increased by sodium fluoride.   The carbonate of sodium appears to produce but little effect upon these.   The silicates are more readily soluble in carbonated water and in dilute solutions of sodium carbonate.   Distilled water seems to act simply as a solvent upon all classes, whereas the other reagents produce more or less decomposition.   Gold is dissolved to some extent in both the silicate and the carbonate of sodium at high temperatures. —Two instruments for the observation of the optical properties of minerals at high temperatures are described by Klein,[3] and a third by Fuess.[4]   One is adapted for use on the microscope, where temperatures not greater than 450° are required.   The second allows of observations at a bright red heat, the source of heat being electrical.   The third is for use with gas.—Miers[5] gives a description of a simple and cheap, though quite accurate goniometer for student's use.—Rinne[6] outlines an easy method of determining the character of the double refraction in uniaxiol and biaxial crystals in converged light, based on the use of the gypsum plate.   Practically the determination depends largely upon the colors of different segments of the microscopic field.   It is especially valuable in determining the sign of weakly doubly refracting substances.—The fifth part of Hintz's Handbuch der Mineralogie[7] concludes the treatment of the mica group, and deals in the usual thorough manner with the chlorite and serpentine groups.

[2]Min. u. Petrog. Mitth., XI, 1890, p. 319.
[3]Neues. Jahrb. f. Min., etc., 1890, I, p. 65.
[4]Ib. B. B., VII, p. 406.
[5]Min. Magazine, IX, 43, p. 214.
[6]Neues. Jahrb. f. Min., etc., 1891, I, p. 21.
[7]Leipzig 1891, p. 641–800.
[1]Ber. d. Deutsch. Chem. Gesell. 1891, p. 1488.

MINERALOGY AND PETROGRAPHY.[1]

Petrographical News.—Several contributions to the subject of the origin of spherulites have recently been made by Messrs. Cross and Iddings, and one on the minerals occurring in hollow spherulites by the latter gentleman and Penfield. Iddings' distinguishes two kinds of spherulites; one composed of radial fibres forming the compact spherulite; and the second consisting of jointed and branching fibres of feldspar, separated by tridymite scales and gas cavities. Gradations between small, dense spherulites composed of micro-felsite, and large ones, the nature of whose structure can be determined, were traced in many instances, and from them the conclusion is reached that microfelsite is in many cases but a microscopic intergrowth of feldspars, elongated parallel to the clino-axis, and quartz, and that the spherulites are but special phases of granophyric growths. The discovery of tourmaline and mica, especially near the margins of spherulites, is an additional proof of the correctness of Iddings's view that spherulites are the result of crystallization of pasty rhyolitic magma under the influence of moisture. These two minerals are younger than the smaller compact radial spherulites of the rock, and older than the final crystallization of the residual magma between the spherulites. In the porous spherulites with branching fibres, or the lithyophysæ, some of the fibres are negative and others positive in the nature of their double refraction; that is, some are orthoclase crystals elongated parallel to *c*, with the plane of the optical axes normal to the plane of symmetry, and others are elongated parallel to *a*, with the plane of the optical axes in the plane of symmetry. The essential characteristic of spherulitic growth is the internal structure of the spherulites. These are not made up of amorphous substances under a strain, but of definitely crystallized minerals arranged radially with one or several centers of crystallization. Under this head, according to the author, would fall granophyric intergrowths, which are radially branching aggregates of orthoclase and quartz. Cross' places emphasis on the valuelessness of the term microfelsite in petrographical nomenclature, as he finds the material to be an aggregate of quartz and orthoclase, two definite minerals, and not the ill defined substance described by Rosenbusch. He attacks both Rosenbusch's and Levy's classification of spherulites as incapable of covering the handsome bodies found by himself in the

[1]Edited by Dr. W. S. Bayley, Colby University, Waterville, Me.
[2]*Bull. Phil. Soc.*, Washington, xi, p. 445.
[3]Ibid, xi, p. 411.

rhyolites of the Silver-Cliff-Rosita mining district in Custer County, Colorado, where spherulites occur of all sizes, up to ten feet in diameter. All are products of the consolidation of a magma, whose composition is

SiO_2	Al_2O_3	Fe_2O_3	FeO	MnO	CaO	MgO	K_2O	Na_2O	H_2O
71.56	13.10	.61	.28	.16	.74	.14	4.06	3.77	5.52

or about ⅔ alkaline feldspar and ⅓ free silica, from which nearly all of the Ca, Mg, etc., had been separated as phenocrysts of plagioclase before the formation of the spherulites. The oldest of the spherulites are minute bodies, in some of which a granophyric growth is detectable. The large ones are found in many generations. Some contain internal cavities, while others are compact. The hollow spherulites are composed of radiating branching orthoclases, with opal and other forms of silica between the fibres, forming a mass through which are scattered minute balls of trydimite or grains of quartz. Another type of spherulite is the trichitic, in which the feldspars are branched and curved to an unusual degree, forming a radiating bunch parallel to whose radii trichites of magnetite are arranged. Both hollow and trichitic spherulites are often surrounded by a supplemental growth in which the feldspar is in very delicate needles. The various generations of spherulites locally make up the entire rock, but usually there is a little residual material consisting of glass, of another radiate growth, or of a combination of both. Compound spherulites are composed of regular orientations of successive growths. The many spherulites of quartz that have been described are thought by the author to be largely feldspar and quartz aggregates, in which the orthoclase is elongated parallel to *c* with the abnormal optical orientation, and thus have a positive double refraction, when they are with difficulty distinguished from quartz microlites. Cross has traced unmistakable prismatic orthoclase down into fibres, and so seems warranted in stating that determinations of the character of the material of spherulites based entirely on the character of the double refraction of the fibres are worthless. Some of the spherulites of the Colorado occurrence consist entirely of positive feldspar, while others are composed of mixtures of this with a negative variety. With reference to the origin of spherulites, Cross reaches the same conclusion as that reached by Iddings; the mass in which spherulitic growth was set up must have come to rest and consequently must have been pasty, since fluidal lines cross the spherulites undisturbed in their courses. During the formation of some of the spherulites the mass again became pasty, and in

certain areas became colloidal, then rapid crystallization was set up
and the branching forms resulted.——Though the main features of
the Rapakiwi granite have long been well-known through the descriptions of Ungern-Sternberg, but little information has been granted
us as to its occurrence and structural peculiarities. A recent article
by Sederholm[4] gives an account of the varieties of the rock-and outlines their modes of occurrence. The peculiarity common to all varieties is the occurrence of porphyritic crystals and the possession of a
granophyric ground mass. The prevailing type possesses phenocrysts
having an elliptical form and surrounded by a rim of oligoclase. The
orthoclase is never pure, but it contains plagioclase particles and grains
of quartz, and biotite or hornblende, the usual constituents of the
groundmass.   These inclusions are often arranged concentrically.
The peculiarity of the structure of the ground mass is the idiomorphism of the quartz, which is often intergrown with the feldspar, lepid-
omelane and hornblende in micropegmatitic forms. The place of the
orthoclase phenocrysts is sometimes taken by an aggregate of orthoclase
and quartz grains, surrounded by a radiating rim of orthoclase and an
exterior one of plagioclase. Miarolitic cavities are filled with fluorite.
As the orthoclase becomes smaller the structure of the rock becomes
more granitic, at the same time the amount of orthoclase decreases
and microcline takes its place. The finest grained varieties occur as
dykes in the others, and are finely granophyric. All these varieties
occur in the Wiborg district in South Finland, where, on account of
their remarkably easy weathering and the consequent production
of granitic debris, they are well known. This easy weathering is
ascribed by the author to mechanical rather than chemical agencies.
At Åland and other regions types are found resembling more or less
closely those described. In some, porphyritic crystals of oligoclase
occur in a micropegmatitic ground mass of quartz and orthoclase, and
in others prophyritic quartzes in a granophyric ground mass. Between
the branches of the quartz in the granophyre are small areas of coarse
grain, and in these are found the miarolitic cavities. Not only do the
rocks described occur in Southern Finland, but they are found also in
the Southwestern portion of the same country, as well as on the islands
off its coast and in the Eastern part of Sweden. All the varieties are
supposed to be phases of the same magma, the coarse-grained, deep-
seated facies and the granophyric surface forms. For granitic rocks
with idiomorphic quartz the author purposes to use the descriptive

[4] Min. u. Petrog. Mitth., xii, p. 1.

term 'anoterite, because probably found at a less depth than the true granite. All the Rapakiwi rocks are thought to be post-archean, but not older than early Cambrian. Structurally they are supposed to represent the source of great radiating dykes and flows (Taphrolites).

——Among some rocks obtained by Doelter from the Cape Verde Islands, Eigel[5] has discovered an augite-diorite, augite-syenite, nepheline basalt, and two doubtful types, which he places respectively with the teschnites and phonolites. The augite-diorite consists of orthoclase in well-formed crystals, and gray plagioclase in lath-shaped individuals, and irregular grains of augite and hornblende. The nepheline basalt occurs as dykes in the diorite, and is composed of phenocrysts of augite and plagioclase, a few olivines and probably orthoclase in a ground mass of nepheline, plagioclase and grains of augite and hornblende. The rocks to which the author assigns a place with the teschnites consist in large part of. a groundmass of altered anorthite and orthoclase, in which are brown augite and hornblende, both altered on their edges to chlorite, a little biotite, magnetite and apatite. The composition of one of these is given as:

SiO_2	Al_2O_3	Fe_2O_3	FeO	CaO	MgO	K_2O	Na_2O	H_2O
39.64	16.98	6.61	9.31	10.58	6.65	3.09	5.95	1.32

The phonolites are fine-grained rocks, made up of porphyritic crystals of acmite and red augite, sometimes zonally intergrown, and hornblende, in a groundmass consisting of microlites of hornblende, augite, biotite, orthoclase, plagioclase, muscovite and magnetite, and little grains of a colorless mineral, probably orthoclase, in an isotropic base. Nepheline could not be detected microscopically, but is supposed to be present as the result of chemical tests. One specimen contains regularly outlined icositetrahedra composed of a nucleus of calcite and sahlite, and an external zone of biotite, that are regarded as altered garnets. Since this rock occurs between a well characterized phonolite and limestone it is thought to be a contact facies of the former.   •

The trachytes, andesites, basalts, etc. of the Upper Eifel have been subjected to a very careful investigation by Vogelsang.[6] The phonolite and the leucite and nepheline basanites have no peculiar characteristics which need be referred to here. The basalts include plagioclase, leucite and nepheline varieties, the former two of which have effected alteration in sandstones and graywackes, with which they are in contact. The trachytes are very much like the Drachenfels rock,

[5] *Mineralog. u. Petrog. Mitth.*, 1890, xii, p. 91.
[6] *Zeit. d. d. geol. Gesell.*, xliii, 1890, p. 1.

and like some specimens of this, it contains tridymite in its ground
mass. The most interesting type studied is hornblende andesite. This
also contains tridymite in its groundmass, and also contains parallel
growths of biotite and hornblende with ⌊OP of the former parallel
to ∞ P∞ of the latter. The hornblende is much corroded, and new
hornblende and feldspar are among the products of its solution.
Large numbers of concretions are characteristic of the rock. These
are granular aggregates of cordierite, andalusite, sillimanite, feldspar,
biotite, pleonast, corundum, rutile, quartz, garnet, zircon and magnetite, and are sometimes schistose. The author thinks that they were
originally inclusions of sillimanite-cordierite gneiss or schist that were
altered by contact with the molten mass of the andesite. He strength-
ens his supposition by treating cordierite-sillimanite rocks with andesite material, when he obtains an abundance of pleonast, which is one
of the most characteristic minerals of the aggregates.——The leucitophyres of the Laacher-See region have again been subjected to a very
thorough microscopical study. Martin[1] has found them to consist
principally of sanidine, leucite, nepheline, augite, and sometimes biotite and melanite phenocrysts in a ground mass of sanidine, nepheline
and green augite, together with a little glassy base. He regards them
as tertiary in age and separates them into two groups, according to
the presence or absence of melanite. The former contain but 48.50—
49.25% of SiO$_2$, while in the latter the percentage of this constituent
rises to 53–54%. The mineral in the rock from Perlenkopf, thought
by Rosenbusch to be perofskite, is melanite. The rocks from Seeberg,
called trachyte by Zickel, are phonolites containing green and violet
augite and nests of olivine. There appear to be gradations between
the phonolites and basanites. The leucite-tufa of the region is a leuci-
tophyre-tuff and the leucitophyre-nepheline tephrites and nephelinites
of the Harmebacher Ley are all nephelinites. Some of the rocks of
Selberg are feldspathic basalts, and nepheline basalts, the latter with
leucite crystals altered to zeolites and augites filled with hornblende
inclusions.

At Democrat Hill and Mt. Robinson in the Rosita Hills, Col.,
are two vents of old solfataras, whose gases have so affected the rhyolite surrounding them that two entirely new and unique rocks have
resulted. At the former place the original rhyolite is now replaced
according to Mr. Cross[2] by a cellular rock composed of alunite and
quartz, and sometimes a little kaolin, whose cavities are lined with

[1]*Zeit. d. d. geol. Gesell.*, xlii, p. 151.
[2]*Amer. Jour. Sci.*, June 1891, p. 466.

crystals of the first two minerals. The alunite is tabular and has a composition :

$$SiO_2 \quad Al_2O_3 \quad K_2O \quad Na_2O \quad SO_3 \quad H_2O \quad Fe_2O_3, etc.$$
$$65.94 \quad 12.95 \quad 2.32 \quad 1.19 \quad 12.47 \quad 4.47 \quad .55$$

At Mt. Robinson the alunite rock is not quite so regular in character, since it contains in places little tablets of barite. Toward the West of the ridge on the top of the mountain is another unique rock composed almost entirely of quartz and diaspore. The composition of this rock is as follows :

$$SiO_2 \quad TiO_2 \quad Al_2O_3 \quad Fe_2O_3 \quad CaO \quad Alk \quad SO_3 \quad P_2O_5 \quad H_2O$$
$$76.22 \quad .11 \quad 19.45 \quad tr. \quad tr. \quad tr. \quad .29 \quad .13 \quad 3.82$$

An analysis of the diaspore crystals implanted on the walls of its cavities yielded $Al_2O_3=83.97$; $H_2O=15.43$.——Among the Archean schists of the Argentine Republic Kühn[9] finds gneisses, mica schists, quartzite and phyllite. In the last three rocks but little of special interest was noted except in the case of the quartzite, where cordierite is supposed to have been discovered. The gneisses are divided into biotitic, muscovitic, and granulitic varieties, and a variety containing two micas. Each group contains fine and medium grained kinds, and one—the biotitic group—embraces a series of "augen-gneiss." The eyes are feldspars, whose outlines indicate that they were originally phenocrysts in a porphyritic granite. Fractured crystal components, peripheral granulation of some of its constituents and an undulous extinction in others, all indicate that the rock has been subjected to enormous pressure. In connection with the discussion as to its origin the author gives an account of the views held on the subject of the origin of gneisses, and discusses their probable correctness. He concludes that gneisses produced by pressure became schistose after their constituents had formed, and that their schistosity is a direct result of the plasticity of the rock mass under pressure, and is not a consequence of numerous fracturings and re-cementings, as Lehmann would have us suppose. With reference to the chemical changes produced by dynamo-metamorphism the author gives descriptions of the alteration of garnet into biotite, and of tourmaline into pinite. He also gives an account of the weathering of garnet into hornblende, and of biotite into chlorite and epidote. The article is particularly interesting in its treatment of the characteristics of schistosity and the origin of the schistose structure.——The constituents of the pegmatite veins

[9] *Neues Jahrb. f. Min.* etc.; B. B. vii, p. 295.

cutting the crystalline schists and granites of the Western part of the Argentine Republic have been carefully examined by Sabensky.[10] They are aggregates of orthoclase, microline, quartz and mica, with plagioclase, biotite, chlorite, tourmaline, garnet, beryl, apatite, zircon and hematite as accessory components. Both the potassium feldspars are intergrown with albite lamellæ, inlaid parallel to a plane between $\infty P_{\overline{\infty}}$ and $2P_{\overline{\infty}}$. The microline offered a fine opportunity for the study of its characteristics. An untwinned specimen gave as a mean of the measurements of its cleavage faces the angles 89° 30.6′ and 90° 29.4′. The peculiar grid-iron structure seen in certain thin sections of the mineral is ascribed to twinning according to the albite law, and not to a combination of twins according to the albite and pericline laws. The arguments brought forward in support of this view are too involved to be dealt with in this place. They are clearly stated in the author's article. Gas and fluid inclusions were formed in the quartz, which mineral often possesses an undulous extinction. Quartz and feldspar are frequently intergrown to give rise to the graphic structure. This is explained by the author as a regular intergrowth of the two minerals in a manner analogous to that of orthoclase and albite, *i. e.*, the quartz follows easy cleavage planes in the feldspar.

Mineralogical News.—Of some rare Argentine minerals recently described by Klockmann[11] the following deserve notice: *Eukarite*, *Umangite* and *Luzonite.* The first named is regarded by the author as a member of the galena group, in spite of the fact that it appears to possess a foliated structure and an hexagonal habit. One analysis yielded: Ag = 43.13 ; Cu = 25.32 : Se = 31.55, corresponding to Ag, Cu, Se, or a jolpaite in which Se replaces S. The mineral occurs in a vein with calcite and umangite, cutting a limestone of unknown age. The *umangite* has heretofore been mistaken for barite. An analysis of the purest material obtainable gave: Cu = 56.03 ; Ag = .49 ; Se = 41.44 ; Co_2, H_2O, etc., = 2.04, which corrected for impurities gives a result corresponding to $Cu_3 Se_2$. Its density is 5.620. The mineral, which is new, is found massive and in the form of a very full-grained granular aggregate. Its hardness is 3. It has a metallic lustre and is opaque. Its streak is black, while its color in consequence of corrosion is a dark, cherry red or violet. The name is taken from the locality in which it occurs—on the West slope of the Sierra de

[10] *Neues. Jahrb. fur Min.*, etc., B. B., vii, p. 359.
[11] *Zeits. f. Kryst.*, xix, 1891, p. 265.

Umango, La Rioja. *Luzonite* was described by Weisbach from Luzon, Philippine Islands, as a substance in all probability isomorphous with famatinite. It however has the composition of enargite, which according to Stelzner is not isomorphous with famatinite. Klockmann thinks the latter mineral and luzonite isomorphous, and regards luzonite as the dimorphous form of enargite. The luzonite is associated with barite as a reddish gray or light-copper-red, massive substance with a hardness of 3.5 and a density 4.390. Its composition is $Cu = 47.36$; $S = 32.40$; $As = 16.94$; $Sb = 3.08$. The locality given for it is Sierra de Famatina, La Rioja.——Analyses of *astrophyllite* from the cryolite locality at St. Peter's Dome, Colo., and of *tscheffkinite* from Bedford Co., Va., yielded Eakins[12] respectively:

Ta_2O_5	SiO_2	TiO_2	LaO_2	$ThO(YEr)_2O_3$	$(LaDi)_2O_3$	Ce_2O_3	Al_2O_3	Fe_2O_3	FeO	MnO
.34	35.23	11.40	1.21				tr.	3.73	29.02	5.52
.08	20.21	18.78	tr(?)	.85	1.82	19.72	20.05	3.60	1.88	6.91

CaO	MgO	K_2O	Na_2O	H_2O	Sp.Gr.	
22	.18	5.42	3 68	4.18		$= R_4'' R_4' Si(SiO_4)_4$
4.05	.55		.06	.94	4.83	

The astrophyllite was very pure, so that the figures of the analysis must be regarded as representing accurately the composition of the substance, especially since they correspond so closely to the formula suggested by Brögger as the result of Bäckstroms investigation. The tscheffkinite was somewhat altered. In thin section Mr. Cross found a brownish transparent amorphous substance crossed by cracks containing reddish brown ochreous decomposition products and bands of colorless minerals that appear to be calcite and sphene, besides several darker minerals. The material analyzed by Price[13] was found upon examination to be as complex in composition, so that it seems probable that the substance has no place among minerals.——The *kamacite*, *tænite* and *plessite* found in the Welland meteorite[14] were so easily separable that Davidson[15] has succeeded in obtaining a sufficient quantity of each for analyses. The kamacite is brittle and of the color of cast iron, while tænite is silvery in lustre and is flexible. The results of the analyses are as follows:

	Fe	Ni	Co	C
Kamacite . .	93.09	6.69	.25	.02
Tænite . .	74.78	24.32	.33	.50

[12]*Amer. Jour. Sci.*, July, 1891, p. 84.
[13]*Amer. Chem. Jour.*, Jan. 1888, p. 88.
[14]*Howell.* *Proc. Rochester Ac. Sci.*, 1890, p. 86.
[15]Ib., p. 64.

Both are magnetic, the latter evincing stronger polarity than the former. In etching, the kamacite is attacked more rapidly than the richer alloy of nickel. Plessite was found to consist of fine lamellae of the two alloys above mentioned.——Brown[16] has carefully examined the *bernardinite* first described by J. M. Silliman,[17] from San Bernardino Co., Cal., as a new mineral resin, and has discovered it to be in all probability the fungus *Polyporus officinalis.*——Weed finds[18] that the ore deposit of the Mount Morgan gold mine in Queensland, Australia, is a siliceous sinter like that of the Yellowstone National Park, impregnated with auriferous hematite. Both the sinter and the hematite are clearly hot spring deposits.——A brief abstract of a paper read by Dr. Foote[19] at the Washington meeting of the A. A. A. S., gives an account of the discovery of black and colorless *diamonds* in a fragment of meteoric iron weighing forty pounds found at Crater Mt., about two hundred miles North of Tucson, Ariz. The diamonds usually occur associated with amorphous carbon in the cavities in the mass, which contains about 3% of Ni.——Census Bulletin No. 49, by Mr. Kunz[20] contains a brief account of the value of gems and precious stones discovered and worked up in the United States during the year 1889. The total value of the materials found within the country amounted to $188,807. Agatized wood, turquoise, zircon, and quartz in the order mentioned are the most important domestic productions falling under the head of precious stones used as ornaments or gems.

[16]*Amer. Jour. Sci.*, July, 1891, p. 46.
[17]Ib., xviii, p. 57.
[18]Ib., Aug., 1891, p. 166.
[19]*Scientific American*, N. Y., Aug. 29, 1891, and *Amer. Jour. Sci.*, Nov., 1891, p.
[20]Census Bulletin No. 49, April 14, 1891.

MINERALOGY AND PETROGRAPHY.[1]

Petrographical News.—Chrustschoff[2] has re-examined the rock of Island Wallamo, in Lake Ladoga, Finland, that was first described by Kutorga as a crystallized labrador-granite. Two varieties were recognized by Chrustschoff, the one a dark-brown dolerite, composed essentially of tabular plagioclase, idiomorphic pyroxene, olivine and a little glass, with sanidine surrounding the plagioclase, and quartz crystals embedded in micro-pegmatite, occurring in the interstices between the plagioclase crystals. The second type is a dark-green diabase-like rock, whose constituents are the same as those of the first-mentioned rock. In this quartz and orthoclase are rare. These rocks are cut by narrow dykes and veins of granophyre, without peculiar features. In explanation of these phenomena the author states that the original rock was an olivine diabase that had solidified, with the exception of its glassy ground mass, when it was intruded by granophyre. The acid magma partially dissolved the crystals and the unsolidified glass of the intruded rock, and so produced an orthoclase quartz aggregate. The basic plagioclase was corroded, and sanidine separated from the mixture formed by its solution in the acid magma, while the remaining acid material cooled as granophyre.——Deecke[3] gives a very detailed account of the geological and petrographical relationships of the gray tufa of the Campagna, Italy, which he believes to be a product of the volcanoes of the Phlegraean Fields. This tufa now consists of a colorless or pale yellow glass, in which are imbedded fragments and crystals of a very soda-rich sanidine, augite and biotite. The rock is thus an augite-trachyte. An analysis of the crystals of sanidine gave:

SiO_2	Al_2O_3	FeO	CaO	MgO	K_2O	Na_2O	Loss	Total
63.79	20.87	1.09	2.06	.41	7.56	3.72	.42	99.92

Besides the fragments of minerals there are also found enclosed in the tufa pieces of augite-trachyte, pumice and obsidian, fragments of hornblende-trachyte, and others of sedimentary rocks. A noticeable and very characteristic feature of the tufa that distinguishes it from others occurring in the same region, are the numerous geodes distributed in great numbers through its mass. These contain a yellowish powder, consisting of sanidine, tufa-fragments and fluorite. They are

[1] Edited by Dr. W. S. Bayley, Colby University, Waterville, Me.

[2] Geol. Fören. i. Stockh. Forh. 13, 1891, p. 149.

[3] Neues. Jahrb. f. Min., etc., 1891, I, p. 286.

supposed to have originated by the gradual decomposition of lapilli
enclosed in the ash.——Hutchings[1] records the existence of tremolite
and garnet in the flags at Shap, England, where they have been altered
by the intrusion of granite through them. The tremolite is produced
in the contact zone within the zone of spotted slates, and the garnet in
about the same zone as the ' knoten,' but in different beds. The minerals producing the spots in the contact rocks are of different natures.
In some cases white mica is the new product found, while in other cases
it is probably andalusite. The clay slate needles that are present in
large quantity in the unaltered flags continue to exist even in those
rocks in which brown mica has begun to form. In the phases in which
brown mica is abundant and newly-formed quartz is present the
needles have disappeared, and in their place are found crystals and
grains of rutile and sphene. In another stage of the contact action
groups of large rutile crystals are observed, and in the neighborhood
of spots are clusters of anatase crystals.——In central Siberia are
mighty dykes and flows of basic rocks, among which Chrustschoff[2]
recognizes ten types of augite—plagioclase—olivine rocks, containing
more or less orthorhombic pyroxene and orthoclase. Their structure
varies from the gabbroitic to the basaltic. Each type is described in
detail and a photograph of it appears with the description. Even in
the most glassy varieties well developed orthoclase exists. The principal structures noted are the gabbroitic, ophitic, with and without
glassy base, anamesitic and aphanitic, with small crystals of feldspar.
——Mr. Rutley[3] describes very briefly a few sections of basalt or andesitic glass from Caradoc Hill, in Shropshire, Eng. At present the rock
contains no olivine, but certain peculiar arrangements of magnetite
grains indicate its former existence in them. With the glass are a
basalt tufa and perlitic felsitic rhyolites with obscure flowage structure, and some with spherulites. An interesting spherulitic and perlitic obsidian is also described by the same author[4] from Pilas, Mexico.
In this the perlitic cracks were certainly formed subsequently to the
spherulites, and were afterwards filled with secondary silica and perhaps other substances.——Much of the so-called anthophyllite and
actinolite in the rocks associated with the iron ores of the Lake Superior region is a monoclinic magnesian amphibole, corresponding to grü-

[1]Geol. Magazine, 1891, p. 459.

[2]Bull. d. l'Acad. Imp. des Sci., d. St. Petersb. Mél. géol. et paleont. T. I, p. 81.

[3]Quart. Jour. Geol. Soc., Nov., 1891, XLVII, p. 534.

[4]Ib. p. 530.

nerite, according to Messrs. Lane and Sharpless.[1] Its refraction is 1.7. It may easily be distinguished from actinolite by its polysynthetic twinning parallel to ∞ P∞ and by its optical characteristics. It is colorless or pale green or brown, and is only faintly pleochroic. An approximately correct analysis gave:

SiO$_2$	Al$_2$O$_3$	Fe$_2$O$_3$	FeO	MgO	(NaK)$_2$O	H$_2$O
76.32	.56	.99	6.96	12.47	tr.	2.80

Fibres of riebeckite or crocidolite have also been discovered by Lane as a secondary growth on the primary hornblende of a syenite from the S. E. ¼ of Sec. 17. T. 49, R. 25, W. in Michigan.——The principal types of olivine and anorthite skeleton crystals in some of the Vesuvian lavas have been well characterized by Rinne[2] in an article illustrated by thirty-eight figures. The olivine skeletons are elongated parallel to the axis *a*. Many are twinned, giving rise to various crosses, in one of which, whose arms intersect at nearly right angles, the twinning plane is ∞ P2̄, a new twinning law for this mineral. Intergrowths of olivine and plagioclase were noted. The anorthite skeletons often show crystallographic faces in grains no larger than .07 mm. in diameter.——Mr. Turner[3] gives a brief account of the geology of Mt. Diablo, in California, describing incidentally a uralitized diabase containing twinned augite, and in some places passing over into a diorite whose hornblende may be secondary, peridotites (lherzolites), pyroxenites (websterite) and gabbros, each of which has given rise to serpentine. In a supplement to Turner's paper, Dr. Melville records the results of the analysis of these rocks together with those of sandstones, shales and a glaucophane schist from the same region. The composition of the schist is as follows:

SiO$_2$	P$_2$O$_5$	Al$_2$O$_3$	Fe$_2$O$_3$	FeO	MnO	CaO	MgO	K$_2$O	Na$_2$O	H$_2$O
47.84	.14	16.88	4.99	5.56	.56	11.15	7.89	.46	3.20	1.98

——A paleozoic leucite-rock, consisting of sanidine, augite, nepheline, and leucite, with the accessories anorthoclase, apatite, zircon and magnetite in a glassy base is mentioned by Chrustschoff[4] from a locality in Russia. The rock is aphanitic and resembles in appearance some of the Hohentwiel phonolites as well microscopically as in microscopic

[1] *Amer. Jour. Sci.*, Dec., 1891, p. 499.

[2] Neues. Jahrb. f. Min., etc., 1891, II, p. 272.

[3] Bull. Geol. Soc. Amer., 2, p. 383.

[4] Neues. Jahrb. f. Min., etc., 1891, II, p. 224.

structure. Sanidine phenocrysts, augite, anorthoclase and leucite lie in a ground mass of sanidine, nepheline and the other above-mentioned constituents.

Mineralogical News.—NEW MINERALS.—*Newtonite* and *Rectorite.*—Messrs. Brackett and J. F. Williams[1] suggest that the kaolinite group of minerals consists of four members, each containing one part of Al_2O_3, two of SiO_2, and one, two, three and four molecules of water respectively. The best-known of these are kaolin, with the composition Al_2O_3 $2SiO_2+2H_2O$, and halloysite, with an additional quantity of loosely combined water. The places of two other members they fill with the new minerals, newtonite and rectorite. The former occurs in lumps in a clay associated with the shales and sandstones of the Barren Coal Measures on Sneed's Creek in Arkansas. It is a pure white, soft, compact, infusible substance, with a density of 2.37. It is only slightly soluble in boiling HCl, but is easily decomposed by hot H_2SO_4 and by boiling NaOH. ·Under high powers of the microscope it appears to form rhombohedrons. Its analysis, calculated for the pure dry material, yielded $SiO_2 = 40.88$; $Al_2O_3 = 35.85$; Loss, $23.27 = Al_2O_3$ $2SiO_2 + 4H_2O$. Rectorite is found in veins in sandstone in the Blue Mountain District, about twenty-four miles North of Hot Springs, Ark. It is in soft white plates, closely resembling mountain leather. Its hardness is less than that of talc, and it is infusible. Upon heating it becomes brittle. The analysis, corrected for impurities, gives $SiO_2 = 54.67$; $Al_2O_3 = 37.22$; Loss $= 8.02$; or, if the excess of silica be regarded as an impurity, $SiO_2 = 49.99$; $Al_2O_3 = 41.16$; $H_2O = 8.84$; besides 8.78% of H_2O at 110°–115°. This corresponds to the first place in the series, viz.: Al_2O_3 $2SiO_2 +$ H_2O + Aq. The index of refraction for the substance is low. It possesses two cleavages inclined to each other, and its acute bisectrix is normal to one of these. ρ . The crystallization is thought to be monoclinic. Tested in the kiln the substance shows properties quite different from those of kaolin.——*Plumboferrite.*—The discovery of this mineral was announced by Igelström as long ago as 1881, but since it has not been noticed in journals outside of Sweden the discoverer[2] reannounces his discovery in a recent article in German. The mineral belongs to the franklinite group, from the other members of which it differs in containing lead in place of zinc or manganese.

[1] *Amer. Jour. Sci.*, July, 1891, p. 11.

[2] *Zeits. f. Kryst,* xix, p. 167.

It is found at Jakobsberg in black, platy masses, with a red streak similar to that of hematite. As usually found it is slightly magnetic, in consequence of the inclusion of impurities. It dissolves easily in HCl, and by H_2SO_4 it is changed to a white mass consisting principally of $Pb SO_4$.

Fe_2O_3	FeO	MnO	PbO	CaO	MgO	$CaCO_3$
55.58	9.83	2.00	21.29	1.55	1.80	7.95

which corrected for the $CaCO_3$ becomes

Fe_2O_3	FeO	MnO	CaO	MgO	PbO
60.38	10.68	2.20	1.67	1.95	23.12

corresponding to $(PbO\ FeO\ MnO)\ Fe_2O_3$.——*Ferro-goslarite* is a zinc sulphate from Webb City, Jasper Co., Mo., of the composition $Zn\ SO_4 = 55.2$; $Fe\ SO_4 = 4.9$; $H_2O = 39.00$; Impur $= .8$. According to Wheeler[1] it occurs as incrustations on the walls of a large body of zinc-blende, with which is associated marcasite and galena. Its formation is due to the oxidation of the zinc and iron sulphides and their subsequent crystallization from solution. It is slightly yellow to brown in color, and is brittle. Its hardness is 2.5, and it loses water on exposure to the air, turning to an opaque, yellow powder in the process. ——*Rowlandite.*—Associated with gadolinite and other yttrium minerals in Llano Co., Texas, Hidden[2] has found a pale drab-green substance that is transparent in thin splinters, and has a density of 4.515. It is easily soluble in acids, leaving a gelatinous residue. Upon alteration it yields a waxy, brick-red product. A partial analysis showed the presence of $SiO_2 = 25.98$; Y_2O_3 etc. $= 61.91$; $FeO = 4.69$; $UO_3 = .40$; $CaO = .19$; Loss $= 2.01$, indicating the formula $R_4'''(SiO_4)_3$.—*Offrétite*[3] is a new zeolite from the basalt of Mt. Simiouse, near Montbrison, in France. It occurs in very small, colorless, hexagonal crystals, with only the base and prism well developed. Their cleavage is basal. Density $= 2.13$ and composition :

SiO_2	Al_2O_3	CaO	K_2O	H_2O	$= (K_2Ca)_2\ Al\ Si_{14} + 17\ H_2O.$
52.47	19.06	2.43	7.72	18.90	

——*Morinite*[4] is a rose-colored mineral with a difficult cleavage par-

[1] *Amer. Jour. Sci.*, March, 1891, p. 212.

[2] *Amer. Jour. Sci.*, Nov., 1891, p. 430.

[3] Gonnard, Bull. Soc. France. d. Min. xiv. p. 58.

[4] Lacroix, Ib., xiv, p. 187.

allel to the triclinic ∞ P$\breve{\infty}$. It is associated with amblygonite at Montebras, Creuse, France. Its crystals contain the three pinacoids, the prisms and several domes. The plane of their optical axis is parallel to ∞ P$\overline{\infty}$, and the extinction in this face is 30°. The optical angle is variable, but it never exceeds 40°. The specific gravity is 2.94, and the mineral contains alumina, soda, phosphoric acid, fluorine and water. With it is associated another hydrated phosphate, crystallizing in pyramids.——*Darapskite, lauterite, iodchromate.*—Dietze[1] describes several new minerals from the Pampas Lantaro in Chile. The first mentioned is a double salt of sodium nitrate and sulphate, of the formula $Na\,NO_3 + Na_2\,SO_4 + H_2O$. It contains $SO_3 = 32.88$; $N_2O_5 = 22.26$; $Na_2O = 38.27$; $H_2O = 7.30$. The mineral is clear and colorless, and is in quadratic tables. Lauterite is particularly interesting as being the first iodate known to occur in nature. Its composition [$I = 64.70$; $CaO = 14.95$] corresponds to $Ca(IO_3)_2$. It occurs in well-developed, large prisms, apparently monoclinic, imbedded in gypsum or implanted in the rocks underlying the pampa. It is transparent and of a yellowish color. Its density is 4.59, and it dissolves quite readily in water. Iodchromate is a peculiar compound in that it is a double salt of the iodate and chromate of calcium, of the formula $7\,Ca(IO_3)_2 + 8\,Ca\,CrO_4$. One analysis gave $I_2O_5 = 58.12$; $CrO_3 = 19.00$; $CaO = 22.01$. The crystals are badly developed. They are of a deep yellow color, and they dissolve easily in water.——*Paramelaconite* and *Footeite.*—Dr. Foote recently obtained from Bisby, Arizona, several specimens affording two new minerals that have been examined by Koenig.[2] One is in bronzy, pyramidal crystals, set in a mass of indigo blue needles, implanted on a mammillary substance composed of a mixture of cuprite and limonite. The dark crystals, named *paramelaconite*, are tetragonal, with a pyramidal habit resembling that of Brazilian anatase, and an axial ratio $a : c = 1 : 1.6643$. The hardness of the new mineral is the same as that of apatite. Its streak is black, and its analysis yielded: $CuO = 87.66$; $Cu_2O = 11.70$; $Fe_2O_3 = .64$. Since the CuO is in such large quantity and since it was found to include little particles of cuprite, the author concludes that the mineral is a dimorph of melaconite. The indigo blue needles are monoclinic combinations of ∞ P∞, ∞ P $\times$ P, P∞ and P$\overline{\infty}$, with an hexagonal habit. Many of the crystals are twins, like those of harmotome. Their composition

[1] *Zeits. f. Kryst*, xix, 1891, p. 447.

[2] *Zeits. f. Kryst*, xix, p. 595.

($CuO = 63.7$; $Cu\ Cl_2 = 13.5$; $H_2O = 22.8$) leads to the formula $8[Cu(HO)_2]CuCl_2 + 4H_2O$. The name chosen for them is Footeite, after their discoverer.——*Rumpfite.*—On the clefts of pinolite, from Jassing, in Obersteiermark, Germany, Firtsch[1] found associated with talc a flaky, greenish-white, translucent mineral with a hardness of 1.5 and a density of 2.675. Under the microscope it appears to be uniaxial and hexagonal. Its composition is:

SiO_2	Al_2O_3	FeO	CaO	MgO	H_2O	Total
30.75	41.66	1.61	.89	12.09	13.12	= 100.12,

corresponding to $H_{26}\ Mg_7\ Al_{16}\ Si_{10}O_{66}$. It deports itself like a chlorite.——*Bolleite.*—Mallard and Cumenge[2] have found among the copper series of Boleo, near Santa Rosalie, in Lower California, a large quantity of a blue copper mineral that often crystallizes in cubes and octohedrons. It is associated with auglesite, cerussite, atacamite and gypsum. An analysis of the cubical forms gave:

Ag	Cu	Pb	Cl	Aq	O(by diff.)
8.80	14.22	49.10	19.48	4.38	4.02,

corresponding to $Pb\ Cl_2 + CuO\ H_2O + \frac{1}{3}\ Ag\ Cl$ or $3[Pb\ Cl(HO).\ Cu\ Cl(HO)] + Ag\ Cl$. The hardness of the crystals is but little superior to that of calcite. Their density $= 5.08$ and their index of rafraction is about 2.07. Sometimes the cubes are modified by the dodecahedron, but the dodecahedral faces have not their usual positions, that one most steeply inclined to each axis occurring nearer the termination of that axis than the one less steeply inclined thereto, so that the edges of the cube are replaced by pairs of two faces making re-entering angles with each other. The optical properties of thin sections of these crystals indicate that they are tetragonal, negative forms that unite to give rise to a pseudo-regular one. The octahedral crystals are in reality tetragonal pyramids with $a : c = 1 : .9873$. Boleite differs from percylite in containing silver.——Sjögren[3] describes three new minerals from Sweden. The first, *astochite*, occurs as a coarse, columnar aggregate of a blue or grayish violet color at Langban. Its cleavage is that of hornblende. Its density varies

[1]Sitzb. Wien, Ak. 99, 1890, p. 1. Ref. Neues. Jahrb. f. Min., etc., 1892, I, p. 31.

[2]Bull. de la Soc. Franc. d. Min. xiv. p. 283.

[3]Geol. Fören. i. Stock. Förh. xiii, 1891, pp. 605 and 781.

between 3.05–3.10, and its extinction between 15°40′ and 17°15′. The analysis of the blue variety yielded:

SiO$_2$	FeO	MnO	MgO	CaO	Na$_2$O	K$_2$O	H$_2$O	F	= Total
56.25	.15	6.49	21.89	5.44	6.17	1.60	1.56	.15	= 99.70

The second, *adelite*, occurs at Langban and at Nordmark in mass of a gray color, with a hardness of 5 and a density of 3.76. Its analysis leads to the formula 2CaO, 2MgO, H$_2$O, As$_2$O$_5$. The third, *svabite*, is probably an apatite. It is found in colorless hexagonal prisms at Harstig Mine, Pajsberg. $a : c = 1 : .7143$. Composition = H$_2$O. 10CaO. 3As$_2$O$_5$.

MINERALOGY AND PETROGRAPHY.[1]

Petrographical News.—Still another attempt to arrive at a just view concerning the chemical relations of eruptive rocks has been made, this time by Lang.[2] Calcium and the alkalies are regarded as the best indicators as to the relationships of the rock masses, and in this respect the new investigation departs widely from older ones, in which the silica was always considered as perhaps the most characteristic of a rock's chemical constituents. After citing a large number of analyses of rocks chosen from carefully examined types of all classes, the author divides rock magmas into four great groups, viz: Those in which the proportion of K_2O present exceeds that of CaO and Na_2O combined, or $K_2O > CaO + Na_2O$, and those in which the proportions of the components correspond to the following formulæ: $Na_2O >$, $CaO + K_2O$, $Na_2O + K_2O > CaO$ and $CaO > K_2O + Na_2O$. Each of these groups is then subdivided into types. In the first group, for instance are two orders in one of which $Na_2O > CaO$, and in the other $CaO > Na_2O$. In the first order fall the Cornwall granites with $CaO : Na_2O : K_2O = 1 : 4 : 14$; the Heidelberg porphyry with an alkali ratio of $1 : 1.5 : 8$; the dyke granite type with a ratio of $1 : 3.7 : 6$, the granite-rhyolite type with a ratio of $1 : 2 : 4$, and the orthophyre type with $1 : 3.8 : 7.3$. The second order includes the Hesse granite, syenite and bolsenite, with the respective alkali ratios $CaO : NaO : K_2O = 2 : 1 : 6$, $2.5 : 1 : 4$ and $1.9 : 1 : 4.8$. The other groups are likewise subdivided into orders, and in each of these are ranged the types. Brief notes accompany the descriptions of each type, and a table giving the percentages of the principal constituents of 247 fresh rocks closes the paper. Some of the relationships brought to light by the author's discussion are so unexpected that it may safely be affirmed that the views put forth in his article will meet with much opposition among petrographers. The granites, for instance, are discovered to occur in different orders, under different groups, the types being often further removed from each other than are normal granite and phonolite.——The new rock iolite, described by Ramsay and Berghell[3] fills the place in Rosenbusch's scheme that was left for the plutonic equivalent of the nephelinites. It is a medium to coarse-grained granular rock forming a large portion of the Mountain Iiwaara in the parish of Kunsamo, Finland.

[1]Edited by Dr. W. S. Bayley, Colby University, Waterville, Me.
[2]Min. u. Petrog. Mitth. xii, p. 199.
[3]Geol. Fören. i. Stockholm Förh. 13, 1891, p. 300.

It consists essentially of nepheline and pyroxene, with iiwaarite (a titaniferous garnet), apatite, sphene and cancrinite. Its structure is allotriomorphically granular, though the pyroxene often possesses one or more crystallographic faces. In the finer grained varieties the garnet is not common, but in the coarse-grained phases of the rock it occurs in large quantity. The pyroxene is zonal, with an almost colorless nucleus, surrounded by six or seven colored zones, in which the extinction is high and the color some shade of green. The mineral occurs either as isolated grains in the nepheline or in little nests of grains in this mineral. Cancrinite is not present in all sections, but in many it is abundant as a decomposition product of nepheline. The relationship of the iolite to nephelinite is shown not only in its possession of a titaniferous garnet, but in its chemical composition as well :

	SiO_2	TiO_2	Al_2O_3	Fe_2O_3	FeO	MnO	CaO	MgO	Na_2O	K_2O	P_2O_5	Loss	Total
Iolite..........	48.79	1.70	19.89	4.39	8.33	.41	11.76	1.87	9.31	1.67	1.70	.99—	98.81
Nephelinite	43.89	1.24	19.25		12.00		10.58	2.81	9.13	1.73	1.39		—102.02

——In consequence of a recent expedition into the Peninsula of Kola, in Northwestern Russia, the senior[1] of the two authors last mentioned has had an opportunity to make a partial geological examination of this little-known territory. He finds the greater portion of the peninsula to be underlain by gneisses, mica schists and Devonian sedimentary beds. The mountains in the neighborhood of Lake Imandra are composed largely of an eleolite syenite, consisting of an intergrowth of albite and microline, eleolite, aegirine, arfvedsonite, eudialite, ainigmatite and a number of other rare and some new species. The aegirine forms long prisms whose extinction is about 4–5° and whose optical angle exceeds 114°. Sometimes a nearly colorless zone surrounds a dark-green kernel, but usually the prisms are dark throughout. An analysis of isolated material gave:

SiO_2	Al_2O_3	Fe_2O_3	FeO	CaO	MnO	MgO	Na_2O	K_2O	Loss
51.82	.60	21.02	8.14	3.01	1.00	1.47	11.87	.85	.50

The arfvedsonite is rare in the normal rock, but is common in its peripheral phases in prismatic grains, whose extinction is 10°30'. The eudialite often possesses an idiomorphic outline bounded by OR, R and ∞ P,. Its double refraction is usually positive, but occasionally some portions of its grains are negative and other portions isotropic.

[1] Fennia. Bull. d. la Soc. d. Géog. de Finlande, 3, No. 7, p. 1.

This phenomenon leads the author to the assumption that eudialite and eukolite are the end members of an isomorphus series, of which the isotropic substances intermingled with the eudialite are intermediate members. Ainigmatite is found only in the peripheral masses as allotriomorphic grains with a pleochroism A = black > B = brown-red > C = carmine. One of the new minerals occurring in the coarse-grained rock has the composition :

SiO_2	Al_2O_3	Fe_2O_3	MnO	CaO	MgO	Na_2O	K_2O	H_2O
55.88	15.19	2.67	9.53	.53	9.06	1.57	6.04	

It is isotropic or weakly doubly refracting. It shows no cleavage, is hard, and has a density of 2.753. Its color is light-red except in certain star-like areas where it is more deeply colored. It is one of the youngest of the rock's components. In the normal rock these constituents are so aggregated as to produce the trachytic structure. In the peripheral varieties aegirine, nepheline and the feldspars are in two generations. These minerals and eudialite occur as phenocrysts in a fine-grained green ground mass of the first three mentioned components, sodalite and the new minerals above referred to. The structure of this aggregate is intermediate between the hypidiomorphic and panidiomorphic. A dyke eleolite syenite from the same region has a thoroughly panidiomorphic ground mass.——In the course of a very exhaustive geological article on Mite Vulture, in Basilicata, Italy, Deecke[1] describes the products of the volcano as lavas and tufas. The former, with the exception of the hauyne-trachyte of Melfi, all possess a similar appearance. They are dark, compact or slaggy rocks with phenocrysts of augite and hauyne in a ground mass of leucite, nepheline, feldspar, augite, biotite, melilite, containing sometimes olivine, garnet, apatite and magnetite. The augite is in well formed idiomorphic crystals, both in the lavas and in the tufas. These are zonal with a yellow augite surrounded by a greenish zone, the material of which also separates as small crystals in the ground mass. The hauyne is the next component in abundance. It possesses the usual characteristics of this mineral, and alters into zeolites, of which the most important is natrolite. The leucite, nepheline, plagioclase and sanidine are usually in such small grains as to be visible only under the microscope. The latter mineral occurs also as an essential constituent in 1 cm. long crystals in the phonolite of LeBraidi, East of Melfi, and in some of the tufas. Olivine is found

[1] Neues, Jahrb. f. Min., etc., B. B. vii, p. 556.

only in the rock of the crater and in bombs, though it was probably more abundant during the first stages of consolidation of nearly all the lavas. Melilite and biotite are also rare, and both seem to have undergone more or less resorption. In addition to the minerals above mentioned bronzite is sometimes found in the olivine inclusions, and natrolite, phillipsite, gypsum, serpentine and kaolin as alteration products of other minerals. The most abundant lava of the volcano is leucitetephrite, with nepheline, leucite, plagioclase and sanidine in varying proportions. A phonolite dyke was discovered at LeBraidi, as already stated. Otherwise phonolitic material is known only as tufa. At Melfi occurs the unique rock, many times described as a melilite hauynophyre. According to the author it should be classed as a hauyne-melilite-nepheline-leucite-tephrite. Basanites cut the older tephritic lavas in the crater of the volcano. Glassy base was detected only in some of the lapilli. The bombs thrown out during the active period of the volcano's history are either olivine bombs or aggregates of augite, biotite and hauyne. In the former the components are olivine, bronzite and biotite, the latter in micropegmatitic intergrowths with the other two. The tufas fall into two classes. In one sanidine and melanite are abundant ; in the other hauyne predominates. The first is the older, and includes the trachyte tuffs of earlier authors. It is a tephrite tufa, which sometimes contains little rounded grains of quartz. The hauyne tufa is more widely spread than the tephritic varieties, and is probably connected genetically with the phonolitic lava.

Mineralogical News—General.—Analyses of $\overset{\circ}{l}angbanite$ having led Flink to the complicated formula $37\,Mn_5\,SiO_8 + 10\,Fe_3\,Sb_2\,O_8$ as expressive of its composition, Backström[1] has thought it worth while to re-examine the mineral in the attempt to learn its true rela-

[1] Zeits. f. Kryst., xix, p. 276.

tionship to other nearly allied species. A new analysis yielded him:

Sb_2O_5	SiO_2	MnO	FeO	CaO	MgO
13.96	9.58	65.44	3.10	1.73	.53

Since chlorine is evolved when the mineral is treated with hydrochloric acid, the author concludes that the manganese is principally in the form of Mn_2O_3, while the remainder of the metal is present as MnO. The conclusion reached is to the effect that langbanite is not isomorphous with any known mineral, but is an isomorphous mixture of Mn SiO$_3$, Mg SiO$_3$ and Mn SbO$_3$.——A large number of *chlorite* analyses are communicated by Ludwig,[1] to whom material was furnished by Teshermak. Among the varieties whose composition was determined are *pennine*, from the Zillerthal, *cronstedtite*, from Pribram, *korundophilite*, from Chester, Mass., *metachlorite*, from Elbingerode, *daphnite*, from Penzance, *tabergite*, from Taberg, *prochlorite*, from the Zillerthal and the Fischerthal, *leuchtenbergite*, from Amity, N. Y., and *clinochlor*, from Achmatowsk, Russia, and from Kariet, Greenland. The figures for korundophilite and leuchtenbergite follow:

	SiO$_2$	Al$_2$O$_3$	Fe$_2$O$_3$	FeO	MgO	H$_2$O	Sp. Gr.
Korund.......... ...	23.84	25.22	2.81	17.06	19.83	11.90	2.87
Leucht...............	30.28	22.13		1.08	34.45	12.61	2.68

These analyses of Ludwig and others that have recently been reported have afforded Tschermak[2] data for the elaboration of a theory concerning the constitution of the chlorites, according to which the members of this group of minerals are regarded as consisting of mixtures of six molecules, four being represented by the known minerals serpentine, amesite, strigorite and chloritoid, and the other two being hypothetical. Dr. Clarke[3] takes exception to Tschermak's views and shows that upon his own theory (that the chlorites are substitution derivatives of normal salts) the composition of these complicated minerals becomes simple, and that his theory is as closely in accord with the facts known as to the structure of the chlorites, as is the theory of the Vienna mineralogist.——In a discussion as to the relations of the recently discovered minerals pinakiolite and trimerite to well known groups Brögger[4] places the former among the rhombic aluminates, and the latter, as a pseudo-hexagonal species, between the olivine and the willemite groups. He further points out the similarity in morphological properties between all of these groups and ascribes their differences to morphotropic action. He regards all the silicates of the general formula R$_4$SiO$_4$ as composing a morphotropic group, in which trimerite is triclinic because R is replaced by two elements, viz: Be and Mn—Several *micas*, *vermiculites*, and *chlorites*[5] have been investigated by Messrs. Clarke and Schneider[6] by the

[1]Min. u. Petrog. Mitth. xii, p. 32.
[2]Akad. d. Wissens. in Wien. , 1891.
[3]Amer. Jour. Sci., xliii, 1892, p. 190.
[4]Zeits. f. Kryst., xviii, p. 877.
[5]Cf. Becke: Min. u. Petrog. Mittl., xi, p. 259.
[6]Amer. Jour. Sci., Sept., 1891, p. 242.

methods[1] already noted in these pages.   Some of the vermiculites seem
to be composed simply of mica molecules, while in others these are
intermingled with molecules possessing the characteristics of those of
chlorite.   Many analyses of vermiculites appear in the paper, and all
of them bear evidence of careful work.——*Calcium-vanado-pyromor-
phite*[2] occurs in the Leadhills, Southern Scotland, as black masses with
an irregular fracture.   It melts easily to a brown crystalline enamel,
and dissolves in hydrochloric acid, leaving a brown residue.   Its density is 6.9–7.0, and in composition it is a pyromorphite with its lead
and phosphorus partly replaced by calcium and vanadium.——

$Pb_3(PO_4)_2$	$Pb_3(VO_4)_2$	$Ca_3(PO_4)_2$	$PbCl_2$	$Cu(OH)_2$	Insol.
52.00	19.20	15.80	11.05	1.50	0.6

Crystallographic.—*Barite* crystals from veins cutting limonite and
siderite, forming lenticular masses in limestone, interstratified with crystalline schists at Huttenberg, Saxony, have been studied by Brunlech-
ner[3].   They are supposed to have originated by the leaching of barium
silicate and its decomposition through the agency of carbonic acid into
barium bicarbonate and silica, and by the further action of iron sulphate upon the barium salt.   Well formed crystals are rare, but the
author has succeeded in detecting upon them twenty-nine forms, of
which the following are new: $\infty P\tilde{\tilde{4}}$, $\infty P2\tilde{2}$, $\infty P30$, $\infty P4\tilde{4}$, $\infty P\tilde{\tilde{4}}$,
$16P\breve{\infty}$, $20P\breve{\infty}$ and $4P\tilde{\tilde{4}}$.——An examination of the crystals of *ullman-
ite* from Sardinia, in the possession of the British Museum, inclines
Miers[4] to regard them as interpenetration twins of tetartohedral forms,
whose apparent holohedral symmetry is due to twinning about the
dodecahedral axis.   If this is so, ullmanite is the first regular tetartohedral mineral known.——Melville[5] has investigated the diaspore crystals in the cavities of the quartz diaspore rock of Mt. Robinson[6].   One
type consists of light-brown transparent forms, elongated parallel to *c*.
Its planes in the order of their development are $\infty P\breve{\infty}$, ∞P, $\infty P\tilde{2}$,
$\infty P\tilde{\tilde{4}}$, $\infty P\tilde{\tilde{4}}$, $P\breve{\infty}$, P and $P\tilde{2}$, with the axial ratio $= .6457 : 1 : 1.0689$.
A second type comprises almost white, opaque crystals with a stout
pyramidal habit, bounded by $\infty P\tilde{2}$, $P\tilde{2}$ and $\infty P\breve{\infty}$.——A very elabor-

[1] AMERICAN NATURALIST, 1891, p. 830.
[2] Collie: Jour. Chem. Soc., lv, 1889, p. 91.
[3] Min. u. Petrog. Mitth. xii, p. 62.
[4] Min. Magazine, ix, p. 211.
[5] Am. Jour. Sci., June, 1891, p. 470.
[6] Cf. AMERICAN NATURALIST, 1892, p. 166.

ate paper on the *vesuvianite* crystals in the serpentine of Testa Ciarva, Alathal, Piedmont, adds but little to our knowledge of these. The results recorded in it but confirm Zepharovich's observations. The crystals examined by the author, Strüver[1], numbered 123, each one of which was carefully measured, both as regards its dimensions and the planes occurring upon it. The number of forms found on each crystal is stated, and the number of crystals upon which each form was observed is mentioned. A plate appended to the article contains thirty-two figures, showing the arrangements of striations and the shapes of elevation and depression figures on the different faces.——In the second part of his article on the symmetry of crystals Beckenkamp[2] gives us some exact information concerning the vicinal planes and the etched figures of the *aragonite* of Bilin and the neighboring localities in Russia, and discusses the polarity of the crystal molecule, with especial reference to the explanation of the electrical properties of the carbonate and of its vicinal planes. The axial ratio of the aragonite crystals examined is .6228 : 1 : .7204.——In a recent article Sohncke[3] explains the structure of circularly polarizing crystals on the basis of his point-system theory of crystal-structure. He succeeds in showing that circularly polarizing crystals may be regarded as composed of thin lamellæ of doubly refracting substance, in which the different layers are revolved a certain number of degrees around their common axis.——Some very complicated twins of feldspar from the Pantelleria rocks are described by Foerstner[4]. They exhibit in the same group combinations of all the principal twinning laws known for the species. ——The *calcites* of fifteen localities in Baden have undergone the same exhaustive examination in the hands of Sansoni[5] as have those of so many other well known occurrences.——Miers[6] has measured the fourth crystal of *krennerite* ($Au_2 Ag_3 Te_8$) reported in mineralogical literature. It is from Nagyag, Hungary, and contains six new forms, viz.: $2P\breve{\infty}$, $3P\breve{\infty}$, $4P\breve{\infty}$, $2P\breve{2}$, $3P\breve{2}$ and $\frac{1}{4}P\breve{2}$. The axial ratio is 1.0651 : 1 : .5388.

Miscellaneous.—Following the work of Clarke and Schneider on the constitution of the natural silicates comes the report of a series of

[1]Neues. Jahrb. f. Min., 1891, p. 1.
[2]Zeits. f. Kryst. xix, 1892, p. 241.
[3]Ib., xix, p. 530.
[4]Ib., xix, p. 560.
[5]Ib., xix, p. 821.
[6]Miner. Magazine, ix, p. 182.

investigations on similar bodies made by Thugutt[1] in Dorpat. The
author recounts the results of his experiments of digesting certain
compounds with water and various chemicals for a long time at a high
temperature, and describes minutely the products formed. By using
the proper ingredients a series of sodalites was produced, in which
sodium silicate, the corresponding selenite, sulphite, chlorate and other
salts take the place of the chloride in the most common sodalite. The
details of the experiments cannot be given, although they are extremely
interesting. The formula thought to represent best the chemical properties of natural sodalite is $4(Na_2O, Al_2O_3, 2SiO_2) + 2\,Na\,Cl$. The
treatment of corundum, a few silicates and natural glasses with water
and alkaline carbonates shows clearly that each reagent is efficient in
hydrating the substances upon which it acts. Many other conclusions
of interest are reached through the author's investigations, but they
cannot be mentioned here for lack of space.

[1] Mineralchemische Studien, Dorpat, 1891, p. 128.

MINERALOGY AND PETROGRAPHY.[1]

Petrographical News.—One of the most valuable contributions to American petrography that has yet appeared is that volume of the Arkansas Geological Survey Report that treats of the eruptive rocks of the State. In it the late Dr. J. F. Williams[2] gives an excellent account of the little-known but very interesting eleolite and leucite rocks that occur as bosses and dykes in Pulaski, Saline, Hotsprings, Garland and Montgomery Counties. It would be well worth the while to give a full abstract of the author's careful investigation of these extremely rare rock-types, but space allows merely a reference to the mere outline of his work. Especial importance is attached to the study of the eleolite syenites at the present time, particularly where its plutonic and dyke forms occur together, since Rosenbusch has recently prophesied the existence of a group of dyke forms which he calls monchiquites, that will be found to occupy a position among the eleolite rocks corresponding to that held by the camptonites, among the plagioclase rocks. The age of the Arkansas eruptives is probably late Cretaceous. In Pulaski County they form the main mass of Fourche Mountain. The most abundant variety here is that locally known as ' blue granite.' It is a granitic porphyritic rock in which the phenocrysts are orthoclase, and the ground mass is a granular aggregate of cryptoperthite, arfvedsonite, diopside, biotite, eleolite and other accessory components. Arfvedsonite is the most prominent dark constituent, while eleolite among the lighter-colored components is scarcely more than an accessory. The rock thus resembles the augite-syenite, called laurvikite by Brögger,[3] but it contains so much more hornblende that Williams decided to give it a new name, pulaskite. Analyses of the rock and its feldspar follow:

	SiO_2	Al_2O_3	Fe_2O_3	FeO	CaO	MgO	MnO	K_2O	Na_2O	H_2O	P_2O_5
Rock.	60.03	20.76	4.01	.75	2.62	.80	tr.	5.48	5.96	.59	.07
Felds.	66.95	17.87	.90		.52	.24		7.82	5.20	.30	

A second rock, occurring along the base of the mountain and a short way up its sides in the form of sheets, is known as ' gray granite.' Its structure is trachytic granular, with large twinned microline-perthites,

[1] Edited by Dr. W. S. Bayley, Colby University, Waterville, Me.
[2] Ann. Rep. Geol. Surv. of Ark. for 1890, vol. ii, 457 pp.
[3] Zeits. f. Kryst., xvi, p. 28.

eleolite, biotite, diopside and aegirine in large quantities. The eleolite frequently changes into analcite. The rock, though quite acid, as the analysis shows, is nevertheless an eleolite syenite, closely similar to laurdalite.[1]

SiO$_2$	Al$_2$O$_3$	Fe$_2$O$_3$	CaO	MgO	K$_2$O	Na$_2$O	H$_2$O
59.70	18.85	4.85	1.34	.68	5.97	6.29	1.88

In and around the mountain are small dykes, filled with 'brown granite,' miarolitic eleolite-syenite and quartz syenite. The brown granite is an eleolite-syenite containing orthoclase, diopside, eleolite and biotite, but no amphibole. In chemical composition it is quite like nordmarkite,[2] but mineralogically it is very different. The miarolitic rocks are panidiomorphically granular. In the eleolite-syenite there is a tendency of the eleolite toward idiomorphic forms when it is not in large quantity. In the quartz syenites the quartz tends toward idiomorphism. An analysis of one of these rocks gave:

SiO$_2$	Al$_2$O$_3$	Fe$_2$O$_3$	MnO	CaO	MgO	K$_2$O	Na$_2$O	H$_2$O
64.63	18.15	3.05	1.00	1.54	.50	4.79	5.80	1.08

Rocks, called by the author border rocks, presumably occurring on the peripheries of dykes and bosses of the eleolite-syenite, contain tabular phenocrysts of sanidine and also sodalite and eleolite in a fine-grained groundmass of eleolite, orthoclase and minute idiomorphic pyroxenes and amphiboles that sometimes shows a fluidal structure. Because of their porphyritic structure and their close relationship with the eleolite-syenite they are classed by the writer among the tinguaites.

The augitic rocks associated with the eleolite-syenites in this region are classed in Rosenbusch's new division of dyke rocks monchiquite. They are all dark in color, and all are characterized by the possession of phenocrysts of augite or of biotite. Those containing olivine are placed with monchiquite, while of the non-olivinitic varieties two new groups are formed; one, called fourchite, possesses phenocrysts of augite; in the other, ouachitite, biotite occurs in large quantity. In the former augite constitutes nearly 75% of the entire rock. Its ground mass is now crystalline, but its structure is thought to be the result of the devitrification of a glassy base. In the amphibole ouachitites the phenocrysts are biotite, hornblende and a few augites in a fine-grained

[1] Ib. p. 33.
[2] Ib., xvi, p. 54.

but originally a glassy groundmass. Augite is still the most prominent bisilicate, but it now constitutes scarcely 20% of the rock mass. The groundmass is composed of minute hornblende and augite crystals, with much magnetite and small, highly refractive grains of what the author supposed to be sphene. The base contains many small, lath-shaped crystals of feldspar. In addition to these two members of the monchiquite group, there is probably a true monchiquite at the south end of Allis Mt., in the same district. This rock is remarkable for its great number of sphene grains and for the common occurrence in it of pseudomorphs of biotite after augite.

At a few places where the contact of the syenite with the surrounding shales may be studied, the latter are found to be much changed and to have developed in them small, irregularly bounded feldspars, a few flakes of biotite, grains of magnetite and plates of hematite.

The sequence of the eruptives is believed to be as follows : Eleolite-syenite and pulaskite (same magma), fourchite, pegmatite and miarolitic dykes.

In the Saline County region the rocks are not very different from those of Fourche Mt. The prevailing gray syenite is coarser in grain than that of Fourche Mt., and its orthoclase crystals (intergrown with albite) tend to become porphyritic. The eleolite is also sometimes in large crystals. A plagioclastic variety of the same rock is slightly more granitic than the orthoclase variety. The pyroxene is aegirite instead of diopside. Pulaskite is absent from this region, but in its place occurs a porphyritic syenite with large orthoclase crystals, crystals of arfvedsonite, biotite and an occasional diopside, imbedded in a fine-grained mosaic of orthoclase and amphibole. This rock also contains spheroidal masses of orthoclase crystals that may be pseudomorphs after leucite. The dyke rocks cutting the eleolite-syenite are grouped into eleolite-syenite pegmatites, aegirite tinguaites and eleolite porphyries. The tinguites are non-eleolitic, while the eleolite porphyries contain large rounded eleolites and aegirites enclosed in a fluidal mass of needles of the last-named mineral. The augitic dykes are much more common in this region than elsewhere. They belong to the monchiquites and the amphibole varieties of this rock. The ground mass of the latter consists of a transparent mass, partly isotropic and partly doubly refracting, in consequence of the development in it of plagioclase and orthoclase laths. Contact rocks are rare, but in one idiomorphic crystals of astrophyllite and aegirine were observed.

The Magnet Cove region is especially interesting, not only because of the fine minerals that occur there, but also because of the great variety of rare rocks found in the neighborhood in great quantity. It is to these latter that the minerals owe their origin. Of the eleolite-syenite occurring here three types are distinguished. One is an eleolite mica syenite, a coarse-grained rock in which eleolite has almost completely replaced orthoclase as the alkaline component. Eleolites, apatites, schorlomites and protovermiculites measuring as much as eight inches in diameter are frequently found imbedded in the decomposed rock. The principal constituents of the fresh rock are allotriomorphic eleolite, biotite crystals, idiomorphic zonal melanites, allotriomorphic schorlomites and large round masses of the same mineral, diopside, crystals of sphene, of ilmenite and of magnetite (the latter giving rise to the lodestone from which the Cove takes its name), pyrite and apatite in large crystals and in needles radiately grouped. Besides these there are also found in the decomposed rock ozarkite, protovermiculite, cancrinite and calcite. The second variety of the syenite is an eleolite garnet syenite made up of a granular mixture of eleolite, melanite and diopside, with small quantities of biotite and the usual accessories. The third type is a miarolitic variety too much weathered to yield good sections. Of the eleolitic dyke rocks one is hypidiomorphically granular, with a tendency to the trachytic structure. It contains a large quantity of orthoclase, eleolite in idiomorphic and allotriomorphic grains, large idiomorphic crystals of aegirine, cancrinite that appears to be in part primary, several other accessories and a number of secondary substances, among which may be mentioned aegirine, fluorite and calcite. A pegmatitic dyke consists of huge microline crystals, beautiful crystals of aegirine, eleolite, eudialite, and other decomposition products of eleolite. This rock is the gangue of many of the sphene, natrolite, brucite, manganopectolite and eucolite crystals obtained in this region. Other dykes in which eleolite occurs in two generations, sometimes with and sometimes without orthoclase in the ground mass, and an eleolite tinguite, in which the phenocrysts are orthoclase are also described. But perhaps the most interesting of the rocks of Magnet Cove are the leucite dyke rocks. Of these the author distinguishes between leucite-syenite and leucite-tinguaite. The former is a hypidiomorphic granular aggregate of a pseudomorph after leucite, eleolite, orthoclase and the basic silicates, diopside and biotite. It is generally connected with eleolite-syenite, and is easily recognized by the large crystals of pseudo-leucite scattered through it. These are imbedded in a ground mass in which may be detected eleolite, black garnets and feldspar. No trace of leucite may be discovered in the large crystals. They are now composed of tabular orthoclases, interspersed with small eleolites and. pyroxenes. Within ·the mass the orthoclase is radial, while on its periphery small tabular crystals have their symmetry planes perpendicular to its surfaces. The ground mass has the structure of an eleolite porphyry. The tinguite occurring as dykes possesses two generations of orthoclase, aegirine, eleolite and the pseudomorphs of leucite, besides many accessory components. It differs from the eleolite-tinguite mainly in the possession of orthoclase in large quantity. Analyses of eleolite-tinguite, leucite-syenite, tinguite and of the pseudo-leucites follow:

	SiO_2	TiO_2	Al_2O_3	Fe_2O_3	FeO	CaO	SrO	MnO	MgO	K_2O	Na_2O	Cl	P_2O_5	Loss
Ele-Tin	53.76	.	23.21	1.27	3.18	2.94	.04		.23	7.01	6.97	.02	tr.	1.71
Leu-Sye[1]	50.96	.52	19.67	7.76		4.38		tr.	.36	6.77	7.67			1.38
Leu-Tin[2]	52.91		19.49	4.78	2.05	2.47	.09	.44	.29	7.88	7.13	.53	tr.	1.19
Pseudo-Leu.	55.06		25.26			.60	tr.		.28	10.34	7.60			1.78

On the contact of the eruptives with the country rock hornstones are formed, but these are so indefinite in their character that but little could be learned from their microscopic study. In the sandstones and limestones through which the eleolite-syenite cuts there are, however, important contact minerals, many of which are of world-wide interest. Among those found in quartz rock and sandstone may be mentioned smoky and milky quartz, arkansite, rutile and hematite. The limestone gave rise to perofskite, magnetite, apatite, phlogopite, vesuvianite and monticellite.

In the potash Sulphur Springs region the eleolite and augite rocks are but little different from those of Magnet Cove. The shale in contact with one of the eleolite-syenite dykes, however, differs somewhat from the contact rocks of other regions. Close to the eruptive the shale has a glassy appearance, and is cut by sheets and masses of pink and white minerals and bands of coarse crystalline calcite. Under the microscope sections are seen to consist of plagioclase, wollastonite, apatite and pyroxene in small interlocking crystals. The white masses cutting it are wollastonite, while the pink ones are xonotlite (hydrated wollastonite), in which a tenth of the CaO has been replaced by Na_2O. The author calls the variety natroxontolite.

Outside of the regions above mentioned aegirite-tinguite dykes occur at Hot Springs and at Hominy Hill, in Sec. 27, T. 1 N. R. 14 W.

The basic dykes outside of the syenite areas have been studied by Kemp, whose report comprises the twelfth chapter of the book under

[1] The leucite-syenite contains .54% of NaCl in addition to the sodium given above, besides a trace of SO_3, and the leucite-tinguaite[2] .52% of SO_3 and .48% of insoluble oxides.

review. Most of them are narrow. In composition they are so closely related to the rocks already mentioned in this abstract that they are regarded as genetically connected with them. The most interesting of the dyke masses are the ouachitites. These are dark in color, and are all porphyritic, with very large phenocrysts of biotite and augite. The groundmass in which these are imbedded is composed of augite, magnetite and glass. An analysis of one of these rocks yielded:

SiO_2	TiO_2	Al_2O_3	Fe_2O_3	FeO	CaO	MgO	K_2O	Na_2O	H_2O	CO_2	P_2O_5	Tot.
36.40	.42	12.94	8.27	4.59	14.46	11.44	3.01	.97	2.36	3.94	1.04	=99.84

This rock, which is so very basic, is a constant associate of the more acid eleolite-syenite. All the basic dykes that have thus far been discovered in the State are mentioned in a table containing 280 entries.

In conclusion the author recapitulates his classification as follows: The eleolite plutonic rocks are called eleolite-syenites. From these are sharply separated the dyke forms, mentioned under eleolite-syenite dyke rocks, which are not porphyritic, but in the variety pulaskite (hornblende eleolite-syenite) are trachytic; the eleolite porphyries, in which eleolite phenocrysts occur; and the tinguites, characterized by porphyritic crystals of orthoclase. Among the latter are the eleolite, leucite and aegirite varieties.

Miscellaneous.—New books, etc.—Mr. Lane[1] has recently published two tables for the use of students in optical mineralogy and petrography. One is devoted to the rock forming minerals. These are divided into three classes, the opaque, the isotropic and the anisotropic groups, and the members of the latter class are subdivided in accordance with the strength of their double refraction, their habit and the character of their extinction. There is no doubt but that with a little practice the student may easily learn to distinguish between the various minerals found in rocks if he will only follow the scheme carefully after reading the explanation accompanying it. In the second table the author gives an excellent resumé of Rosenbusch's classification of massive rocks, as elaborated in his ' Massige Gesteine.'——An even more recent tabulation of igneous rocks is that of Dr. Adams[2], since it reflects Rosenbusch's present attitude with respect to rock classification. In it the author represents in a very concise and simple form the differences between the various types of massive rocks, and

[1] Am. Geol., June, 1891, p. 341.
[2] Can. Rec. of Sci., Dec., 1891, p. 463.

suggests at a glance their relationships. In each of three horizontal columns are placed the plutonic, dyke and volcanic rocks in such a way that the corresponding members of all classes fall in vertical columns. The novel features of the classification are the following: Leucite-syenite is made a sub-group under the eleolite-syenites, and the tinguaites are represented as their corresponding granitic dyke forms. The rocks in which leucite, nepheline and melilite replace feldspar are given columns between the orthoclase and plagioclase rocks, as better indicating their chemical affinities than is the case when they are placed beyond the plagioclastic rocks. Each of the three groups is divided in accordance with the presence or absence of olivine, and in the niche for the plutonic nepheline combination is placed the new . type iolite, recently described by Ramsay[1]. The lamprophyric dyke melilite rock is alnoite. Among the lamprophyric plagioclase nepheline rocks are fourchite and monchiquite, the latter with and the former without olivine. Malchite is a new rock described by Osann. It is the granitic dyke form of diorite. The diabases are put in the paleovolcanic class, which, by the way, is not sharply defined from the neovolcanic class. The basic non-feldspathic rocks are separated into the pyroxenites and the peridotites, which names sufficiently define themselves. Dr. Adams deserves the thanks of all petrographers for his enterprise in the preparation of this, the only modern classification of massive rocks published.——INSTRUMENTS.—Salomon[2] describes a simple piece of apparatus by means of which the density of a heavy liquid may be rapidly determined without the inconvenience of the use of a Mohr's balance. It consists essentially of a W-shaped graduated tube, whose manipulation depends upon the principle that the heights of the liquid columns in its two arms will vary with their differences in density. By the aid of the instrument the specific gravity of each mineral in a rock powder may be determined without once emptying it of its solution.——Four new microscopes for crystallographic and petrographical purposes have lately been introduced to the favor of investigators. The three[3] from the manufactory of Zeiss in Jena present no peculiar features. The fourth was made by Nachet to the order of Wyrouboff[4], especially for observations at high tem-

[1]Cf. AMERICAN NATURALIST, 1892, p. 834. By mistake iolite occurs in the wrong column in the table.
[2]Neues. Jahrb. f. Min., etc., 1891, ii, p. 214.
[3]Czapski, Neues. Jahrb. f. Min., etc., B B. vii, p. 497.
[4]Bull. Soc. Franc., d. Min. xiv., 1891, p. 198.

peratures. The objective is below the object, and the nicol prism and the illuminating mirror are above it. The stage is fixed, while the microscope body revolves. Attempts to use converged light with this instrument have failed, because of the great heat to which the condensers are subjected.

MINERALOGY AND PETROGRAPHY.[1]

Quartz and Feldspar Inclusions in Diabase.—Backström[2] has discovered in several Scandinavian diabases inclusions of quartz and feldspar, and has carefully studied the effects produced by the reactions between them and the enclosing rock magma. The diabase consists of labradorite, three augites, magnetite and several secondary substances, and has a structure that differs slightly from the diabasic structure, in that the angles between the plagioclase and the more or less columnar augite, are filled with a groundmass of feldspar laths, pyroxene needles, magnetite and chlorite. Near the quartz inclusions the quantity of the groundmass increases and the diabasic structure becomes obscure, until it finally disappears, and in its place is seen a porphyritic aggregate with thick tabular crystals of plagioclase, well formed augite prisms and large grains of magnetite in a groundmass that occupies a third or a half of the field of view. Very near the quartz the plagioclase become smaller, and spherulites of quartz and feldspar more abundant, of which the latter mineral is either orthoclase or oligoclase. The feldspathic inclusions in the rock are orthoclase, microline and plagioclase. The action of the magma upon the microline is shown in the existence in it of 'solution-spaces,' which are spaces dissolved from the midst of the mineral and afterwards filled with rock magma, from which have crystallized pyrite, magnetite, ilmenite, needles of pyroxene, lath-shaped crystals of oligoclase, grains of quartz, calcite and masses of chlorite. The orthoclase inclusions sometimes become granulated, or filled with long lenticular areas of a feldspar whose origin is the same as that of the minerals in the solution spaces. Plagioclase fragments have also undergone granulation, and the new feldspar has crossed the original twinning planes irrespective of their directions. Other fragments are more completely penetrated by the magma, so that everywhere throughout their mass may be detected small areas of the diabasic groundmass with its spherulites. The peripheries of the grains are often marked by growths of new, colorless, transparent plagioclase, crystallographically continuous with the enclosed mineral. Mica, hornblende and other iron-bearing compounds seem less capable of resisting the action of the

[1]Edited by Dr. W. S. Bayley, Colby University, Waterville, Me.
[2]Bihang t. k. Svensk. Vet. Akad. Handl., B. 16, Afd. II, No. 1.

magma than the feldspar fragments. They are consequently often entirely dissolved, leaving behind them scarcely a trace of their former existence. When cores of these remain they are surrounded by magnetite, quartz (?) and sometimes newly formed biotite. Before closing his article the author calls attention to the differences in the action of diabasic and basaltic magmas upon enclosed fragments, the most noticeable being that in the former case no glass inclusions are developed in the material of the enclosed substances, while in the latter case these are abundantly formed.

The Basalts of Cassel.—Among the basaltic rocks occurring in the neighborhood of Cassel, Fromm[1] distinguishes limburgites, basalts and nepheline-basalts. All are porphyritic, with olivine, augite and labradorite (in the plagioclastic varieties) as phenocrysts in a groundmass of augite, plagioclase, probably sanidine, magnetite, ilmenite, hematite, mica and apatite in the feldspathic rocks and in the nepheline basalts with a groundmass of much the same character except that nepheline replaces the feldspar. Besides, a little glassy base is always present. Some of the olivine is brecciated as if broken by the movement of the rock in which it occurs. Again it is corroded, when it is often surrounded by a rim of glass, which the author thinks is due to rapid cooling of the dissolved portion of the mineral. The augites are often zonal, with the deeply colored zones within in the plagioclase basalts, and in the nepheline varieties with the lighter colored zones in the interior. The nepheline present in the nepheline basalts is either in the form of a uniformly distributed groundmass in which are imbedded microlites of augite or in little nests between the other minerals, or finally as small areas with crystal cross sections. Mellilite was detected in a nepheline basalt and good plates of pleochroic ilmenite were observed in several plagioclase basalts. Glass was present in the limburgite, in several plagioclase basalts and in one nephelinic variety. Quartz fragments included in the plagioclase basalts are surrounded by rims of augite crystals, between which and the nucleus is often a zone of glass. Sometimes the quartz has entirely disappeared, leaving only glass encircled by a crown of augite. Analyses of all the varieties of the rocks studied accompany the paper.

The Rocks of the Piedmont Plateau.—In an excellent article on the structure of the Piedmont Plateau in Maryland, G. H. Williams[2] divides the region between the Catoctin Mountains and the

[1]Zeits. d. d. Geol. Ges. xliii, p. 48.
[2]Bull. Geol. Soc. Amer., Vol. ii, p. 301.

Costal area into two parts, a Western one underlain by fragmental rocks in which but slight alteration has been effected, and an Eastern one characterized by both sedimentary and eruptive rocks that have been strongly metamorphosed by pressure. In the first area phyllites, including sericite, chlorite and ottrelite schists, sandstones made up of undoubtedly clastic grains, and a few crystalline limestones occur. In the Eastern area the piedmont rocks are gneisses, some of which are evidently eruptive and others probably sedimentary; quartzites and quartzite-schists, in which all evidences of fragmental origin have disappeared, coarse crystalline marbles containing phlogopite, tremolite, scapolite, etc., and acid and basic eruptives. Each of these classes, except those belonging to the eruptive division, is briefly characterized, and pictures[1] of the sandstone of the Western area and of the quartzite of the Eastern area are given for comparison.

The Diorite of the Andes, first mentioned by Stelzner as a characteristic rock of this mountain range, has been closely studied by Möricke[2] in the occurrence at St. Cristobal near Santiago in Chile. It is closely associated with andesites, both hornblendic and augitic, with propylite and tufas, in such a way that the author is led to look upon them all as facies of the same rock mass, the diorite representing the deep-seated phases and the andesites the surface flows. The hornblende, augite and hypersthene-augite, andesites and the diorites have the usual characteristics of these rocks. With respect to structure an intermediate phase occurs in the propylite, which consists of phenocrysts of plagioclase and green hornblende in a ground mass of plagioclase, chlorite and epidote. The propylite is much altered, while the other rocks are fresh. The geological relations of the different rock types correspond with the conclusions outlined as above. The 'stocks' of diorite, often forming the peaks of the range, are the denuded cones of old massifs.

The Porphyry of Monte Doja.—Pelikan[3] has re-examined the rock discovered by Suess at Monte Doja, in the Adamello group of mountains, and described by him as a reddish brown porphyry. The structure is porphyritic and the groundmass is dark brownish red in color. Under the microscope the thin section shows a colorless base containing yellowish-brown wisps of biotite, thin prisms and fine

[1] C. R. Keyes, Ib. p. 321.
[2] Min. u. Petrog. Mitth. xii, p. 143.
[3] Min. u. Petrog. Mitth., xii, p. 156.

needles of rutile and six-sided prisms of tourmaline. The colorless mass in which these are imbedded contains plagioclase, orthoclase and quartz. The porphyritic crystals in this groundmass are cordierite. They all possess a more or less hexagonal cross section, and the usual optical properties of the mineral. Its inclusions are biotite plates and rutile, and tourmaline crystals like those of the groundmass. The composition of the rock is represented by the figures:

SiO_2 TiO_2 Fe_2O_3 Al_2O_3 FeO CaO MgO K_2O Na_2O H_2O
56.88   2.66   20.86   4.54   1.29   3.15   7.48   .90   2.36

Corresponding to 18% of cordierite, 20% of mica, 30% of orthoclase and 13% of plagioclase. The origin of the rock is unknown.

Wernerite Rocks occurring at several places in the Pyrenees are stated by Lacroix[1] to be altered feldspathic ǀeruptives. The original form of the varieties from Sallix and Ponzac was a hornblende diabase containing olivine, biotite and sphene. The borders of the feldspar, which is labradorite, are often fringed by little plates of dipyr. Veinlets of these penetrate the plagioclase until in some cases all the labradorite is replaced by large plates of the scapolite, many of which are of much larger size than the feldspars from which they were formed. They moreover possess a uniform orientation over large areas, so that the structure of the rock passes from the microlitic to the granular. When dynamic agencies have modified it broken pieces of the scapolite are often found. This indicates to the author that in these cases the alteration of the feldspar preceded the crushing. On the other hand, broken pieces of all the other minerals are sometimes found cemented by unbroken dipyr, in which case the formation of the latter mineral was subsequent to the shattering. The alteration is thus a strictly chemical process. The paper closes with a reference to the Norwegian scapolite rocks and their comparison wth those of the Pyrenees.

Petrographical News.—The phonolites[2] of Montusclat and of Lardeyrols in the Ardennes, in France, contain small crystals and crystallites of lävenite, the latter often grouped into forms resembling the skeletons of large crystals. The larger crystals are golden yellow, with a strong pleochroism.

[1]Bull. Soc. Franc., d. Min. xiv, p. 16.
[2]Lacroix. Ib., xiv, p. 15.

Anatase and brookite are described[1] as occurring in French rocks, the former in the mica porphyrites of Pranal, Puy-de-Dom, and the latter in the chloritized mica of the limestone of Ville-es-Martin, in the Loire-Inférieure, and in the mica of a mica porphyrite from Pouchon, in Cercie, and in the granite of Lacourt, Ariège.

The same author[2] announces the discovery of octahedra of cristobalite associated with tridymite in a piece of quartz inclusion from the basalt of Mayen in Rhenish Prussia.

Nova Scotian Gmelinite.—A careful analysis of *gmelinite* from the Five Islands, Nova Scotia, has been made by Pirsson,[3] and the properties of the mineral have been investigated. The mineral occurs in seams in a decomposed trap. In thin sections its crystals are found to be composed of a colorless compact outer zone enclosing a flesh-colored, friable inner nucleus. Their habit is short pyramidal, and their axial ratio $a : c = 1 : .7345$. Two methods of twinning were observed, viz., interpenetration twinning with oP the twinning plane, and contact twinning parallel to ½R. The double refraction is weak and negative, with $\omega-\epsilon$ for sodium light $= .0033$. Density $= 2.037$, and composition :

SiO_2	Al_2O_3	Fe_2O_3	CaO	K_2O	Na_2O	H_2O
50.35	18.33	.26	1.01	.15	9.76	20.23

The author regards the mineral as a sodium-chabazite, in which $Na : Ca = 8 : 1$. The sodium is thought to exert a marked morphotropic action, since the crystallographic planes of the gmelinite cannot be conveniently referred to the axes of chabazite. It may, therefore, be considered as a distinct mineral species.

Crystallography of Cerussite, etc.—Pirsson[4] also contributes a few crystallographic notes on *cerussite, gypsum* and *krämmerite.* Twinned crystals of the first-named mineral from the Red Cloud Mine, Ariz., are arrow-shaped, with the plane ∞ P3 the twinning plane. The gypsums are from Girgenti, Sicily. They are twinned according to the usual law, but ½P∞ so largely predominates over other forms that they resemble basal planes and cause the twinned crystals to resemble in symmetry an orthorhombic form. The krämmerite studied is from

[1] Lacroix. Ib., xiv p. 191.
[2] Ib., xiv, p. 185.
[3] Amer. Jour. Sci., July, 1891, p. 56.
[4] Amer. Jour. Sci., Nov., 1891, xlii, p. 405.

Texas. Pa., and is the same as that examined by Cooke[1] in 1867. Three new forms were observed on it, viz.: ∞P2, ⅓P2 and ⅓P. The prism ⅓P2 is very characteristic and is seldom wanting. The author also corrects the axial ratio for mordenite to $a : b : c = .401 : 1 : .4279$. β = ... and briefly describes skeleton crystals of *hematite* from Durango, Mex., filled with cassiterite and pseudomorphs of the latter mineral after the former.

Mineralogical News.—Dr. Hoffman[2] mentions a quartzite from the North shore of St. Joseph Island, in Lake Huron, whose joint planes are coated with limonite containing numerous tiny spherules composed of a nucleus of silica, coated with a humus-like substance, which in turn is overlain by a layer of metallic iron. The spherules form 33.54 of the mixture. They have a density of 6.8612 and the following composition: Fe = 88.00; Mn = .51; Ni = .10; Co = ...

of the joint cracks. P and oP are the only faces observed on the anatase.——Crystalline masses of *ilvaite* are reported by Hoffman[1] from a twenty-foot vein near the head of Barclay Sound, Vancouver Island, British Columbia. Its hardness is 5.5, density 3.85 and composition :

SiO_2	Al_2O_3	Fe_2O_3	FeO	MnO	CaO	MgO	H_2O
29.81	.16	18.89	32.50	2.22	13.82	.30	1.62

Celestite from Scharfenberg, near Meissen, Saxony, according to Stuber[2] occurs in druses of little blue crystals whose axial ratio is .7807 : 1 : 1.2834, and whose habit is determined by the planes $P\breve{\infty}$, oP and $\frac{1}{2}P\overline{\infty}$. Another druse is made up of brown or wine-yellow crystals whose habit differs from that of the blue crystals in that oP is small. Their axial ratio differs also, since $a : b : c = .7834 : 1 : 1.2962$.

Weed and Pirsson[3] describe orthorhombic crystals of *sulphur* from around the vents at the Highland Hot Springs and at Crater Hills in the Yellowstone National Park, and stalactitic and fibrous growths of orpiment and realgar in crevices of the rock surrounding vents in the Western part of the Morris Geyser Basin in the same territory.

Nesquehonite exists in the anthracite mine of Mure, Isère. Its analysis by Friedel[4] gives results that prove its identity with the nesquehonite of Genth and Penfield.

Syntheses.—While heating mica with soda and sodium sulphate in a tube the mixture became dry, leaving a residue in which the Messrs. Friedel[5] discovered little rhombohedra, with the composition of *nosean*. Upon heating mica, soda and sodium carbonate for two days at 500° the same experimenters obtained little hexagonal pyramids of a substance with the composition $3 (2 SiO_2, Al_2O_3, Na_2O) + Na_2 CO_3 + 2H_2O$, which the authors regard as a calcareous rich cancrinite. If amorphous alumina be heated at 530° in a solution of soda in a closed tube the excess of the first material will separate[6] upon cooling as *corundum*. If heated at 400° *diaspore* is produced. When the soda contains a little aluminum and calcium carbonates, *calcite* crystals

[1] Ib., Nov., '91, p. 482.
[2] Zeits. f. Kryst., xix, 1891, p. 437.
[3] Ib., Nov., 1891, p. 401.
[4] Bull. Soc. Franc. d. Min., xiv, p. 60.
[5] Bull. Soc. Franc. d. Min., xiv, p. 69.
[6] G. Friedel. Ib., xiv, p. 7.

Texas, Pa., and is the same as that examined by Cooke[1] in 1867. Three new forms were observed on it, viz.: ∞ P2, ½P2 and ½P. The prism ½P2 is very characteristic and is seldom wanting. The author also corrects the axial ratio for *mordenite* to $a : b : c = .401 : 1 : .4279$. $\beta = 88° 29' 46''$, and briefly describes skeleton crystals of *hematite* from Durango, Mex., filled with cassiterite and pseudomorphs of the latter mineral after the former.

Mineralogical News.—Dr. Hoffman[2] mentions a quartzite from the North shore of St. Joseph Island, in Lake Huron, whose joint planes are coated with limonite containing numerous tiny spherules composed of a nucleus of silica, coated with a humus-like substance, which in turn is overlain by a layer of metallic iron. The spherules form 58.85% of the mixture. They have a density of 6.8612 and the following composition: $Fe = 88.00$; $Mn = .51$; $Ni = .10$; $Co = .21$; $Cu = .09$; $S = .12$; $P = .96$; $C = ?$; insol. res. $= 9.76$. The insoluble residue consisted of little concentric bodies with the composition: $SiO_2 = 93.95$; $Al_2O_3 = 1.13$; $Fe_2O_3 = 1.02$; $CaO = .62$; $Mg = .31$; Loss $= 2.97$.

Analyses of uraninite from two new localities are given in a recent paper by Hillebrand[3]. The first is that of a specimen obtained from Marietta, Greenville Co., S. C., and the other that of one from the Villeneuve Mine, Ottawa, Quebec. Neither of the two was pure.

	UO_3	UO_2	ThO_2	ZrO_2	CeO_2	(La. etc.)	Yt. etc.	CaO	PbO	H_2O
S. C.....	83.95	1.65	.20	.19	2.05	6.16	.41	3.58	und.
Que.....41.06	34.67	6.41	?	.40	1.11	2.57	.39	11.27	1.47

	N	SiO_2	Insol.	Fe_2O_3	X
S. C.................	undet	.20		tr.	tr.
Que...................	.86	.19	.13	.10	.09

Results of the analyses of the same mineral from Llano Co., Texas, and from Johann-georgenstadt, Sax., show the presence of nitrogen in each.

Williams[4] has discovered *anatase* crystals in a jointed slate from five miles South of Bremo Bluffs, Buckingham Co., Va. The crystals are small and are associated with pyrite and quartz, which cover the faces

[1] Ib., 1867, xliv, p. 201.
[2] Amer. Geol., viii, p. 105.
[3] Amer. Jour. Sci., Nov., 1891, p. 391.
[4] Ib., Nov., '91, p. 431.

of the joint cracks. P and oP are the only faces observed on the anatase.——Crystalline masses of *ilvaite* are reported by Hoffman[1] from a twenty-foot vein near the head of Barclay Sound, Vancouver Island, British Columbia. Its hardness is 5.5, density 3.85 and composition :

SiO$_2$	Al$_2$O$_3$	Fe$_2$O$_3$	FeO	MnO	CaO	MgO	H$_2$O
29.81	.16	18.89	32.50	2.22	13.82	.30	1.62

Celestite from Scharfenberg, near Meissen, Saxony, according to Stuber[2] occurs in druses of little blue crystals whose axial ratio is .7807 : 1 : 1.2834, and whose habit is determined by the planes $P\widetilde{\infty}$, oP and $\frac{1}{4}P\widetilde{\infty}$. Another druse is made up of brown or wine-yellow crystals whose habit differs from that of the blue crystals in that oP is small. Their axial ratio differs also, since $a : b : c = .7834 : 1 : 1.2962$.

Weed and Pirsson[3] describe orthorhombic crystals of *sulphur* from around the vents at the Highland Hot Springs and at Crater Hills in the Yellowstone National Park, and stalactitic and fibrous growths of orpiment and realgar in crevices of the rock surrounding vents in the Western part of the Morris Geyser Basin in the same territory.

Nesquehonite exists in the anthracite mine of Mure, Isère. Its analysis by Friedel[4] gives results that prove its identity with the nesquehonite of Genth and Penfield.

Syntheses.—While heating mica with soda and sodium sulphate in a tube the mixture became dry, leaving a residue in which the Messrs. Friedel[5] discovered little rhombohedra, with the composition of *nosean.* Upon heating mica, soda and sodium carbonate for two days at 500° the same experimenters obtained little hexagonal pyramids of a substance with the composition 3 (2 SiO$_2$, Al$_2$O$_3$, Na$_2$O) + Na$_2$ CO$_3$ + 2H$_2$O, which the authors regard as a calcareous rich cancrinite. If amorphous alumina be heated at 530° in a solution of soda in a closed tube the excess of the first material will separate[6] upon cooling as *corundum.* If heated at 400° *diaspore* is produced. When the soda contains a little aluminum and calcium carbonates, *calcite* crystals

[1] Ib., Nov., '91, p. 432.
[2] Zeits. f. Kryst., xix, 1891, p. 437.
[3] Ib., Nov., 1891, p. 401.
[4] Bull. Soc. Franc. d. Min., xiv, p. 60.
[5] Bull. Soc. Franc. d. Min., xiv, p. 69.
[6] G. Friedel. Ib., xiv, p. 7.

Texas, Pa., and is the same as that examined by Cooke[1] in 1867. Three new forms were observed on it, viz.: ∞ P2, ½P2 and ½P. The prism ½P2 is very characteristic and is seldom wanting. The author also corrects the axial ratio for *mordenite* to $a : b : c = .401 : 1 : .4279$. $\beta = 88° 29' 46''$, and briefly describes skeleton crystals of *hematite* from Durango, Mex., filled with cassiterite and pseudomorphs of the latter mineral after the former.

Mineralogical News.—Dr. Hoffman[2] mentions a quartzite from the North shore of St. Joseph Island, in Lake Huron, whose joint planes are coated with limonite containing numerous tiny spherules composed of a nucleus of silica, coated with a humus-like substance, which in turn is overlain by a layer of metallic iron. The spherules form 58.85% of the mixture. They have a density of 6.8612 and the following composition: $Fe = 88.00$; $Mn = .51$; $Ni = .10$; $Co = .21$; $Cu = .09$; $S = .12$; $P = .96$; $C = ?$; insol. res. $= 9.76$. The insoluble residue consisted of little concentric bodies with the composition: $SiO_2 = 93.95$; $Al_2O_3 = 1.13$; $Fe_2O_3 = 1.02$; $CaO = .62$; $Mg = .31$; Loss $= 2.97$.

Analyses of uraninite from two new localities are given in a recent paper by Hillebrand[3]. The first is that of a specimen obtained from Marietta, Greenville Co., S. C., and the other that of one from the Villeneuve Mine, Ottawa, Quebec. Neither of the two was pure.

	UO_3	UO_2	ThO_2	ZrO_2	CeO_2	(La. etc.)	Yt. etc.	CaO	PbO	H_2O
S. C.....	83.95	1.65	.20	.19		2.05	6.16	.41	3.58	und.
Que.....41.06		34.67	6.41	?	.40	1.11	2.57		.39	11.27 1.47

	N	SiO_2	Insol.	Fe_2O_3	X
S. C..................	undet		.20	tr.	tr.
Que..................	.86	.19	.13	.10	.09

Results of the analyses of the same mineral from Llano Co., Texas, and from Johann-georgenstadt, Sax., show the presence of nitrogen in each.

Williams[4] has discovered *anatase* crystals in a jointed slate from five miles South of Bremo Bluffs, Buckingham Co., Va. The crystals are small and are associated with pyrite and quartz, which cover the faces

[1]Ib., 1867, xliv, p. 201.
[2]Amer. Geol., viii, p. 105.
[3]Amer. Jour. Sci., Nov., 1891, p. 391.
[4]Ib., Nov., '91, p. 431.

of the joint cracks. P and oP are the only faces observed on the anatase.——Crystalline masses of *ilvaite* are reported by Hoffman[1] from a twenty-foot vein near the head of Barclay Sound, Vancouver Island, British Columbia. Its hardness is 5.5, density 3.85 and composition :

SiO_2	Al_2O_3	Fe_2O_3	FeO	MnO	CaO	MgO	H_2O
29.81	.16	18.89	32.50	2.22	13.82	.30	1.62

Celestite from Scharfenberg, near Meissen, Saxony, according to Stuber[2] occurs in druses of little blue crystals whose axial ratio is .7807 : 1 : 1.2834, and whose habit is determined by the planes $P\breve{\infty}$, oP and $\frac{1}{4}P\overline{\infty}$. Another druse is made up of brown or wine-yellow crystals whose habit differs from that of the blue crystals in that oP is small. Their axial ratio differs also, since $a : b : c = .7834 : 1 : 1.2962$.

Weed and Pirsson[3] describe orthorhombic crystals of *sulphur* from around the vents at the Highland Hot Springs and at Crater Hills in the Yellowstone National Park, and stalactitic and fibrous growths of orpiment and realgar in crevices of the rock surrounding vents in the Western part of the Morris Geyser Basin in the same territory.

Nesquehonite exists in the anthracite mine of Mure, Isère. Its analysis by Friedel[4] gives results that prove its identity with the nesquehonite of Genth and Penfield.

Syntheses.—While heating mica with soda and sodium sulphate in a tube the mixture became dry, leaving a residue in which the Messrs. Friedel[5] discovered little rhombohedra, with the composition of *nosean*. Upon heating mica, soda and sodium carbonate for two days at 500° the same experimenters obtained little hexagonal pyramids of a substance with the composition $3 (2 SiO_2, Al_2O_3, Na_2O) + Na_2CO_3 + 2H_2O$, which the authors regard as a calcareous rich cancrinite. If amorphous alumina be heated at 530° in a solution of soda in a closed tube the excess of the first material will separate[6] upon cooling as *corundum*. If heated at 400° *diaspore* is produced. When the soda contains a little aluminum and calcium carbonates, *calcite* crystals

[1] Ib., Nov., '91, p. 432.
[2] Zeits. f. Kryst., xix, 1891, p. 437.
[3] Ib., Nov., 1891, p. 401.
[4] Bull. Soc. Franc. d. Min., xiv, p. 60.
[5] Bull. Soc. Franc. d. Min., xiv, p. 69.
[6] G. Friedel. Ib., xiv, p. 7.

Texas, Pa., and is the same as that examined by Cooke[1] in 1867. Three new forms were observed on it, viz.: ∞P2, ⅓P2 and ⅓P. The prism ⅓P2 is very characteristic and is seldom wanting. The author also corrects the axial ratio for *mordenite* to $a : b : c = .401 : 1 : .4279$. $\beta = 88° 29' 46''$, and briefly describes skeleton crystals of *hematite* from Durango, Mex., filled with cassiterite and pseudomorphs of the latter mineral after the former.

Mineralogical News.—Dr. Hoffman[2] mentions a quartzite from the North shore of St. Joseph Island, in Lake Huron, whose joint planes are coated with limonite containing numerous tiny spherules composed of a nucleus of silica, coated with a humus-like substance, which in turn is overlain by a layer of metallic iron. The spherules form 58.85% of the mixture. They have a density of 6.8612 and the following composition: Fe = 88.00; Mn = .51; Ni = .10; Co = .21; Cu = .09; S = .12; P = .96; C = ?; insol. res. = 9.76. The insoluble residue consisted of little concentric bodies with the composition: SiO_2 = 93.95; Al_2O_3 = 1.13; Fe_2O_3 = 1.02; CaO = .62; Mg = .31; Loss = 2.97.

Analyses of uraninite from two new localities are given in a recent paper by Hillebrand[3]. The first is that of a specimen obtained from Marietta, Greenville Co., S. C., and the other that of one from the Villeneuve Mine, Ottawa, Quebec. Neither of the two was pure.

	UO_3	UO_2	ThO_2	ZrO_2	CeO_2	(La. etc.)	Yt. etc.	CaO	PbO	H_2O
S. C.....		83.95	1.65	.20	.19	2.05	6.16	.41	3.58 und.	
Que.....	41.06	34.67	6.41	?	.40	1.11	2.57	.39	11.27	1.47

	N	SiO_2	Insol.	Fe_2O_3	X
S. C..................	undet	.20		tr.	tr.
Que...................	.86	.19	.13	.10	.09

Results of the analyses of the same mineral from Llano Co., Texas, and from Johann-georgenstadt, Sax., show the presence of nitrogen in each.

Williams[4] has discovered *anatase* crystals in a jointed slate from five miles South of Bremo Bluffs, Buckingham Co., Va. The crystals are small and are associated with pyrite and quartz, which cover the faces

[1]Ib., 1867, xliv, p. 201.
[2]Amer. Geol., viii, p. 105.
[3]Amer. Jour. Sci., Nov., 1891, p. 391.
[4]Ib., Nov., '91, p. 431.

of the joint cracks. P and oP are the only faces observed on the anatase.——Crystalline masses of *ilvaite* are reported by Hoffman[1] from a twenty-foot vein near the head of Barclay Sound, Vancouver Island, British Columbia. Its hardness is 5.5, density 3.85 and composition :

SiO_2	Al_2O_3	Fe_2O_3	FeO	MnO	CaO	MgO	H_2O
29.81	.16	18.89	32.50	2.22	13.82	.30	1.62

Celestite from Scharfenberg, near Meissen, Saxony, according to Stuber[2] occurs in druses of little blue crystals whose axial ratio is .7807 : 1 : 1.2834, and whose habit is determined by the planes $P\widetilde{\infty}$, oP and $\frac{1}{2}P\overline{\infty}$. Another druse is made up of brown or wine-yellow crystals whose habit differs from that of the blue crystals in that oP is small. Their axial ratio differs also, since $a : b : c = .7834 : 1 : 1.2962$.

Weed and Pirsson[3] describe orthorhombic crystals of *sulphur* from around the vents at the Highland Hot Springs and at Crater Hills in the Yellowstone National Park, and stalactitic and fibrous growths of orpiment and realgar in crevices of the rock surrounding vents in the Western part of the Morris Geyser Basin in the same territory.

Nesquehonite exists in the anthracite mine of Mure, Isère. Its analysis by Friedel[4] gives results that prove its identity with the nesquehonite of Genth and Penfield.

Syntheses.—While heating mica with soda and sodium sulphate in a tube the mixture became dry, leaving a residue in which the Messrs. Friedel[5] discovered little rhombohedra, with the composition of *nosean*. Upon heating mica, soda and sodium carbonate for two days at 500° the same experimenters obtained little hexagonal pyramids of a substance with the composition $3 (2 SiO_2, Al_2O_3, Na_2O) + Na_2 CO_3 + 2H_2O$, which the authors regard as a calcareous rich cancrinite. If amorphous alumina be heated at 530° in a solution of soda in a closed tube the excess of the first material will separate[6] upon cooling as *corundum*. If heated at 400° *diaspore* is produced. When the soda contains a little aluminum and calcium carbonates, *calcite* crystals

[1] Ib., Nov., '91, p. 432.
[2] Zeits. f. Kryst., xix, 1891, p. 437.
[3] Ib., Nov., 1891, p. 401.
[4] Bull. Soc. Franc. d. Min., xiv, p. 60.
[5] Bull. Soc. Franc. d. Min., xiv, p. 69.
[6] G. Friedel. Ib., xiv, p. 7.

accompany the corundum. These experiments throw some light upon the production of corundum in metamorphosed limestones. Under similar conditions magnesia yields *brucite.* When calcium silicate is treated with sodium borate at high temperatures under pressure *datholite* crystals are formed, according to Wyrouboff.[1]

Miscellaneous.—A new contribution to the discussion of the cause of optical anomalies is from the pen of Karnojitzky[2], who ascribes the phenomena to polymerism. Paramorphic substances are thought to be polymeric, with the higher polymer more stable than the simpler one. The latter usually possesses a higher degree of symmetry than the former, and hence, when it forms first it gives a higher degree of symmetry to its crystals than is possessed by those of its paramorph— the more stable, more complicated compound. The author develops this idea in a very logical and clear manner, and instances many examples to indicate the probability of the correctness of his statements.

Retgers[3] proposes to test the isomorphism of different substances by forming of them mixed crystals, which, if the two substances used be isomorphous, will differ continuously in their physical properties. In the case of colored salts the test most easily applicable is that of color. If potassium chlorate and the corresponding permanganate be dissolved in water and a drop of each be placed in an object glass and allowed to come in contact, the crystals formed at the junction of the two drops will be intermediate in color between those formed at a distance from it. In the case of salts that are not isomorphous no gradation in color will be observed. In his article the author mentions the many compounds that he has discovered to be isomorphous.

Lemberg[4] suggests a modification of micro-chemical analysis by which minerals are detected rather than their constituent elements. For instance, instead of distinguishing between hauyne and sodalite by studying their elements, the author would study the effect of the minerals themselves upon reagents, e. g. sodalite will precipitate silver chloride from a weak nitric acid solution of silver nitrate, whereas hauyne will not do so. Chabazite reacts rapidly with ammonium chloride and generates ammonia, while thomsonite, analcite and leucite

[1] Ib., xiv, p. 197.
[2] Zeit. f. Kryst., xix, p. 571.
[3] Zeit. Phys. Chem., viii, July, 1891, p.   , and Jour. Chem. Soc., lx, p. 1151,
[4] Lemberg Die Krystallanalyse, p. 57.
[5] d. Geol. Gesele., xlii, 1891, p. 737.

are without effect upon it of any kind. The principal minerals studied by the author that yield definite and distinctive results are hauyne, sodalite, scapolite, chabazite, calcite, witherite, cerussite and anglesite and others of less importance. The method of manipulation necessary to produce good results is described in each case with great minuteness.

Dufet[1] describes a method for the determination of the comparative values of the indices of refraction by means of the prism and total reflection, and Lavenir[2] gives an account of a new process by which the optical orientation of any crystal may be determined.

A report on "The Mineral Resources of the Province of Quebec," by Ells[3], contains a history of the various mining industries of the district. In the same annual report Hoffmann[4] publishes a list of the minerals occurring in Canada, and Ingalls[5] gives statistics relating to the production and exportation of the mineral products of Canada.

[1]Bull. Soc. Franc. d. Min., xiv, p. 130.
[2]Ib., xiv, p. 100.
[3]Pt K. Ann. Rep. Can. Geol Survey, 1888–89.
[4]Pt. I. Ib.
[5]Pt. S. Ib.

MINERALOGY AND PETROGRAPHY.[1]

The Basalt of Stempel.—Bauer's[2] description of the basalt of
Stempel, near Marburg, and its concretions and inclusions is one of
the most excellent pieces of petrographical work that has appeared in
a long time. A favorable opportunity has enabled the author to secure
a splendid suite of specimens of this rock so noted for its beautiful
zeolites. It consists of the usual constituents of basalt, viz.: plagioclase, augite and olivine in a groundmass of augite and feldspar
microlites in a base of glass. The plagioclase is andesine without
peculiar characteristics. The augite is also without special features
except that it is frequently zonally developed, with a dark-green kernel and brown-colored coats, in which the extinction decreases from 48°
to 36°. The olivine is so well bounded by crystal planes that the relations of the shapes of the cross-sections to the crystallographic axes
have been well worked out. Twins parallel to $P\breve{\infty}$ are not uncommon.
The liquid inclusions, upon careful study, are found to differ from
those of the olivine of the concretions (Knollen), and the glass inclusions are learned to have a different composition from the glass forming the groundmass of the rock. One of the most interesting features
of the rock is the occurrence of amygdaloidal cavities, coated within
by a layer of glass, whose limits are sharply defined. Sometimes a
partition of this glass divides a cavity into two, and occasionally several concentric partitions give rise to a series of chambers that are
strikingly like the chambers in Idding's lithophysæ. The olivine
bombs included in the rock consist largely of bronzite and chrome-
diopside grains cemented by olivine substance. The bronzite is present in two varieties, one an almost opaque greenish-brown kind, and
the other a transparent olive-green variety. Picotite is also present
quite abundantly in grains and aggregates of grains in most of the
bombs. The effect of the action of the rock magma upon its inclusions is seen in the granulation of the pyroxenes, and the effect of the
material of the bombs upon the magma is shown in the presence of microlites of hypersthene in the veins of the rock that ramify the bombs.
Since the minerals of the bombs contain characteristic inclusions not
common to lherzolitic rocks, and since, moreover, the olivine and bronzite are sometimes found in forms never seen in lherzolite, the author

[1] Edited by Dr. W. S. Bayley, Colby University, Waterville, Me.
[2] Neues Jahrb. f. Min., etc., 1891, ii, p. 156.

concludes that these bodies are not inclusions torn from a deep-seated basic rock as is sometimes thought, but that they are concretions of the basic minerals of the basalt, formed during the intratellurial period of its magma's history.   Another interesting feature of the Stempel·occurrence is the abundance and variety of true inclusions found therein.   These are limestone, quartz, feldspar and amphibolite fragments and others torn from a cordierite rock.   The limestone has produced but little effect upon the surrounding rock other than rendering its texture coarser by increasing the size of its feldspathic constituents. The limestone itself has suffered little change.   The quartz fragments are all surrounded by rims of green augite crystals, and in their interior they are filled with swarms of cavities either empty or filled with liquid.   Sandstone inclusions now consist of grains of quartz, cemented by a glass that has originated in the fusion of the cement of the original rock.   This glass sometimes contains trichites and magnetite grains, when it is colorless; sometimes it is devoid of them and is colored brown.   The glass cement also frequently contains drops of glass that differ from the enclosing material in that it dissolves readily in hydrochloric acid, while the latter is unaffected by this reagent.   The included substance is regarded as the pure glass produced by the solution of the cement of the sandstone, while the insoluble variety is that to which silica has been added by the corrosion of the quartz grains. The finer grained sandstones have yielded basalt-jasper.   In their glassy constituent are numerous crystals of apatite that are similar in most of their properties with the nepheline and ·cordierite crystals observed by Zirkel in some of the basalt-jaspers described by him. The orthoclase inclusions are penetrated by tiny veins of glass.   Both the feldspar and the glass contain small violet octahedra of some spinel and blue pleochroic needles of glaucophane, while tridymite plates occur in the latter substance.   An aggregate of orthoclase and plagioclase contains flecks of green glass between the grains that is thought to be fused mica, while the feldspar is filled with sillimanite needles.   The other inclusions present features that are worthy of notice, but they cannot be described in the present place.   The article will well repay the reader for its perusal.

The Crystalline Rocks of Tammela, Finland.—The archean rocks in the vicinity of Tammela, in the South-western part of Finland, are crystalline schists, granites, gabbros, porphyrite and vitrophyres. A gray granite, Sederholm[1] thinks, is closely related to the

[1]Min. u. Petrog. Mitth., xii, p. 97.

gabbros and diorites of the region, which appear as though basic separations from the same magma as that yielding the granite. The most abundant rock is a muscovite granite. Next in importance is a uralite-porphyrite, whose uralitic phenocrysts are complete pseudomorphs of augite. All the constituents of the rock show much alteration. The plagioclase is changed to epidote and zoisite, and between the secondary products of this mineral are newly formed plagioclase and hornblende, and in addition there are frequently accumulations of biotite, whose form leads to the supposition that they are pseudomorphs after olivine. In its original condition this rock was probably a basalt. A plagioclase-porphyrite, an amygdaloid and glassy rocks with the composition of an acid basalt also occur in the region. Tufas accompanied the outflow of basalt, but in this as in the other rocks described the character of the original substance has been greatly obscured by alteration. In discussing the cause of the chemical changes that have been effected, the author ascribes the most powerful action to water in connection with pressure. Many of the rocks show evidences of dynamo-metamorphism. A schistosity has been superinduced in nearly all of the types, but the crushing and breaking of grains that are such striking phenomena in most instances of this kind, are here absent. The pressure exerted its influence principally in increasing the solvent power of the water. Very little change in the chemical composition of the rocks has resulted from the alteration, in spite of the fact that their mineralogical composition has been totally changed.

Petrographical Notes.—The breccias and porphyries of Pilot Knob, Mo., have repeatedly been stated to be metamorphised fragmentals. Haworth[1] has examined their relations to other rocks and has carefully studied their thin sections with the result that they are pronounced by him true eruptives, the latter, quartz-porphyries, exhibiting flowage structure, and other evidences of having once been liquid, and the former, porphyry breccias, with fragments of porphyry cemented by a groundmass that was once a fluid volcanic lava.—— Cordierite-bearing chiastolite schists are briefly mentioned by Klemm[2] as forming part of the contact belt of the Lausitz granite at Dubring, and dykes of hornblende-porphyrite as cutting the granite at this place and at Schmerlitz, in Saxony.——In a brief communication Kemp[3] speaks of the existence of several dykes of a very much altered

[1] Bull. No. 5, Geol. Surv. of Mo., p. 5.
[2] Zeits. d. d. Geol. Ges., xliii, 1891, p. 526.
[3] Amer. Jour. Sci., Nov., 1891, p. 410.

peridotite in the Portage sandstones near Ithaca, N. Y.——In a horn-blende-andesite inclusion in the Capucin trachyte Lacroix[1] finds one cavity containing magnetite, biotite, fayalite and hypersthene—a different association of minerals from that in any other cavity. The most interesting of these minerals is the fayalite, which occurs in tiny crystals with a golden yellow color, due to a ferruginous pigment.

Mineralogical News.—A series of new analyses of *amarantite* from the Mina de la Campania, near Sierra Corda, Chile, give: $SO_3 =$ 35.46; $Fe_2O_3 = 37.46$; $K_2O = .11$; $Na_2O = .59$; $H_2O = 28.29$, corresponding to $Fe_2 S_2O_9 + 7H_2O$. The mineral has a specific gravity of 2.286, and at 110° it loses three molecules of water. Its axial ratio as determined by Penfield[2] is $a : b : c = .7692 : 1 : .5738$ with $a = 95° 38' 16''$; $\beta = 90° 23' 42''$, $\gamma = 97° 13' 4''$, and $2Ena = 63° 3'$. In sections parallel to the trachy-pinacoid the extinction is 16°–17° in acute β. *Sideronatrite* from the same place occurs in fine orange or straw-yellow fibres, with orthorhombic symmetry (not monoclinic as Raimondi asserts). Its density is 2.355. A mean of several analyses yielded:

$$\begin{array}{cccc} SO_3 & Fe_2O_3 & Na_2O & H_2O \\ 44.22 & 21.77 & 16.39 & 17.07 \end{array} = 2Na_2 SO_4 Fe_2 S_2O_9 + 7H_2O.$$

The mineral suffers a loss of four molecules of water at 110°. Associated with sideronatrite are little white masses composed of a substance with hexagonal optical properties. It is positive with $\omega =$ 1.558, $\epsilon = 1.613$ for yellow light. Its density = 2.547–2.578, and its composition is: $H_2O = 11.89$; $SO_3 = 51.30$; $Fe_2O_3 = 17.30$: $Na_2O = 19.63$; $K_2O = $ ca. .16, corresponding to $3Na_2 SO_4. Fe_2(SO_4)_3 + 6H_2O$. With these analyses are also given those of a *picropharmacolite* from Joplin, Mo., of *pitticite* from the Clarissa Mine, Utah, of *gibbsite* from Chester Co., Pa., and of *atacamite* from Chile. The analysis of the first mentioned mineral leads to the formula $(H_2 Ca Mg)_3 As_2 O_8 + 6H_2O$. The pitticite gave: $H_2O = 17.64$; $As_2O_5 = 39.65$; $Fe_2O_3 = 33.89 = 4Fe_2(As O_4)_3. Fe_2(OH)_6 + 20H_2O$. The mineral is not a mixture of the sulphate and arsenate of iron as is the German variety. No definite conclusion was reached as to the composition of the gibbsite other than that it is a hydrous aluminum phosphate.——Though *columbite* has been known to exist in the Black

[1] Bull. Soc. Franc., d. Min., xiv, p. 10.

[2] Zeits. f. Kryst., xviii, p. 585.

Hills in Dakota for some six years past, the first accurate account of
its occurrence and of its composition has but just been communicated
by Mr. Headden.[1] The mineral together with *tantalite* is often present in the stream tin of the hills. It is also found imbedded in beryl
at the Etta Mine and associated with other minerals at the various
other mines in the district. Fourteen analyses of crystals obtained
from the different localities are given. Some of these correspond with
the formula $3R\ Cb_2O_6 + 2R\ Ta_2O_6$, with $R = Fe_{½}\ Mn_{½}$. As the density of the mineral becomes greater the proportion of tantalum to
columbium increases, passing from 1 : 6 to 1 : 1½; thus indicating
that columbite and tantalite are isomorphous substances. Analyses
follow: I. Turkey Creek, Col.; II. Yolo Mine, S. Dak.; III. Tantalite, associated with stream tin at the Grizzly Bear Gulch, S. Dak.;
IV. Manganiferous columbite, from Advance Claim, 1½ miles S. of
Etta Mine.

	Cb_2O_5	Ta_2O_5	SnO_2	WO_3	FeO	MnO	CaO	Sp. Gr.
I.	73.45	2.74	.21	1.14	11.32	9.70	.61	5.383
II.	24.40	57.60	.41		14.46	2.55	.73	6.592
III.	3.57	82.23	.32		12.67	1.33		8.200
IV.	47.22	34.27	.32		1.89	16.25		6.170

Mr. Headden's results are interesting as indicating the widespread
occurrence of these two rare minerals in the Black Hills region, and
his paper is valuable for the great number of analyses contained in it.
——Laspeyres[2] has reexamined the *saynite* (of V. Kobell) from Grube
Grüneau, in Kirchen on the Sieg, in Germany, where the mineral
occurs in crystals. He finds it to be a mixture of polydymite with
other sulphides, as he declared it to be some time since. *Ullmanite*
crystals from Siegen, in the same neighborhood, are described as consisting of cubes with striations parallel to the pyritoid edge, or of
cubes, dodecahedrons and octahedrons combined with more complicated forms, among which are many parallel hemihedral ones. Its crystallization thus corresponds with that of the Sardinian Ullmanite
described by Klein.[3] A rare chance was also afforded Laspeyres for
the study of the crystallization of *wolfsbergite*, from Wolfsberg, in the
Harz. The new crystals obtained by him are tabular parallel to oP,

[1] Amer. Jour. Sci., Feb., 1891, p. 89.
[2] Zeits. f. Kryst, xix, 1891, p. 417.
[3] Neus. Jahrb. f. Min. etc., 1883, i, p. 180 and 1887, ii, p. 169.

and have the macro-zone more highly developed than the brachy-zone.
They show clearly that Groth is correct in regarding the mineral as
isomorphous with amplectite, scleroclase and zincenite. The axial
ratio, calculated from pyramidal faces that gave good reflections, is
$a : b : c = .5283 : 1 : .6234.$——The little-known members of the
mesotype group on the Puy-de-Dôm have recently been described by
Gonnard[1] in some detail as regards localities. An analysis of the natrolite from the Puy-de-Maman yielded: $SiO_2 = 48.03$; $Al_2O_3 =$
26.68; $Na_2O = 15.61$; $H_2O = 9.62$; and that of the Tour de Gev-
illat gave: $SiO_2 = 47.88$; $Al_2O_3 = 26.12$; $Na_2O = 15.63$; $CaO =$
$.45$; $H_2O = 9.80.$——The same author[2] has made a crystallographic
study of the *barites* of the Puy-de-Dôm. All crystals of this substance
are beautifully modified but none show new forms. A peculiarly habited *aragonite*[3] from the Neussargnes Tunnel, Cantal, contains the new
forms $\frac{1}{2}P\breve{\infty}$ and $\frac{1}{4}P.$——The investigation of the nature of the nitrogen found in *uraninite*, promised some time ago, has been continued
by Hillebrand[4] without, however, very great success. The most careful analyses of specimens from Glastonbury, Ct., and from Arendal,
Norway, yield respectively:

UO_3	UO_2	ThO_2 etc.	PbO	CaO	H_2O	N	Fe_2O_3	SiO_2	Insol.	Sp. Gr.
23.03	59.93	11.10	3.08	.11	.43	2.41	.29	.16	.89	9.622
26.50	44.18	13.87	10.95	.61	undet.	1.24	.24	.50	1.19	

The principal result of the analyses is to the effect that all uraninite
contains more or less nitrogen, sometimes amounting to as much as
$2\frac{1}{2}\%$. The condition in which the element exists is unknown, but it is
probably different from any hitherto observed in the mineral kingdom.
Another result indicated is that the formulas that have been accepted
as expressing the composition of the mineral do not do so. Specimens
from many of the classical localities have been analyzed, and in nearly
every case errors have been detected in the original analyses. The
author concludes that while uraninite in general contains the same
constituents, it varies widely in composition, and its physical characteristics are often as distinct as are the chemical differences.——The
keramohalite from Pico de Teyde, in the Canary Isles, is in little imper-
fectly developed crystals imbedded in a yellowish white hygroscopic

[1]Bull. Soc. Franç. d. Min., 1891, xiv, p. 165.

[2]Ib., xiv, p. 174.

[3]Ib., xiv, p. 183.

Geol. Survey, No. 78, p. 43.

granular mass, in the neighborhood of solfataras. The soluble substance extracted from this mass by Hof[1] gave:

SO_3	Al_2O_3	Fe_2O_3	FeO	CaO	MgO	Na_2O	H_2O
38.62	13.96	.94	.66	.22	.04	2.37	42.01

The form of the crystals as determined by Becke[2] is tabular parallel to $\infty P\bar{\infty}$. They have a weak negative double refraction. The axis of mean elasticity is inclined 48° to $\infty P\bar{\infty}$, and that of the least elasticity 13° to $+P\bar{\infty}$. The crystallization is monoclinic with $a : b : c = 1 : ? : .825 \ \beta = 97° \ 34'$.——In the druses of a massive *garnet* used as a flux in the copper smelters at Kedobek, Caucasia, are found crystals of garnet that rival in beauty the famous Tyrol varieties. They are bounded by the forms 2O2, ∞O and occasionally 3O⅓, and all the faces are brilliant. Their color is wine to honey-yellow and their composition[3] is represented by:

SiO_2	CaO	Al_2O_3	Fe_2O_3	Loss	
39.12	35.84	22.73	1.76	.15	$=Ca_3 \ Al_2(SiO_4)_3.$

——According to Branner[4] inexhaustible beds of *beauxite* occur near Little Rock and Benton, Ark., that are supposed to be genetically related in some way with eruptive granites. The material is pisolitic in structure. The composition of one variety as shown by a partial analysis is:

Al_2O_3	SiO_2	Fe_2O_3	TiO_2	Loss
55.64	10.38	1.95	3.50	27.62

——The handsome *calcite*[5] twins from Guanajuato, Mexico, that have been known for some time, are usually the scalenohedron R[3], twinned parallel to —⅓R. Corresponding pairs of faces on each individual are so developed that their combination has a monoclinic habit, resembling strongly the swallow-tailed twins of gypsum. The forms recognized in the crystals are mentioned in the paper and six figures accompany it.

[1]Min. u. Petrog., Mitth. xii, p. 39.

[2]Ib., p. 45.

[3]Müller : Neues. Jahrb. f. Min., etc., 1891, i, p. 272.

[4]Amer. Geologist, vii, 1891, p. 181.

[5]Pirsson: Amer. Jour. Sci., Jan., 1891, p. 61.

——Frenzel[1] has made a new analysis of *gordaite* and has found it to be identical with ferronatrite, while Arzruni has examined its crystals and declares them to be rhombohedral with $a : c = 1 : 55278$.——C. Schneider[2] gives good analyses of six basaltic hornblendes, all of which contain over 4% of TiO_2.

[1]Zeits. f. Kryst., xviii, p. 595.
[2]Zeits. f. Kryst., xviii, p. 579.

MINERALOGY AND PETROGRAPHY.[1]

The Eruptives of Cabo-de-Gata.—The eruptive rocks of the Cabo-de-Gata region in southeastern Spain are pumiceous, glassy and granular liparites, andesites, dacites and an occasional basanite. The liparites are rare as fragments in a liparitic tufa and as small dykes cutting the fragmental rocks. The dacites cover a large stretch of country. They are the most abundant types in the region, and are developed in great variety. Two principal groups are distinguished. The first is characterized by the abundance of its phenocrysts, among which are large hornblendes, and by the possession of augite and hypersthene. Augite occurs in their groundmass, quartz is scarce, and their feldspar is almost exclusively plagioclase. In the second group phenocrysts are less common. Biotite is the predominant colored constituent. Quartz and sanidine are both plentiful and the rock thus verges toward the liparites. All the components of these dacites have been very minutely described by Osann.[2] But few of them present special peculiarities. The most interesting features connected with them are the alteration of hornblende into pyroxene and the intergrowth of augite and bronzite, with the pinacoids and prisms of the two minerals parallel. The andesites, which are best developed in the southern and southeastern parts of the region, are hornblendic and biotitic varieties. A mica andesite from the Rambla del Esparto contains an enormous number of granular inclusions composed of cordierite (?) biotite, spinel, sillimanite, corundum, andalusite, plagioclase, rutile, zircon, garnet, quartz and apatite. They are regarded as having resulted from the metamorphism of blocks brought from below, and the crystallization of andesite components upon them. The spinel occurs in dark-green and grayish-red crystals, the former sometimes surrounding the latter. A dacite from Mazarron contains phenocrysts of cordierite,[3] whose prismatic crystals often reach a length of 1 cm. The forms

---

[1] Edited by Dr. W. S. Bayley, Colby University, Waterville, Me.

[2] Zeits. d. Deutsch. Geol. Ges. 1891, xliii, p. 688.

[3] Cf. AMERICAN NATURALIST, 1890, p. 69.

observed on them are ∞P, $\infty P\breve{\infty}$, $\infty P\overline{\infty}$, oP, P, and $\frac{1}{2}P$. In the neighborhood of ore veins the mineral is changed into pinite.

A Melilite Rock from North America.—From the bed of the Ottawa River, near Ste. Anne, not far from Montreal, Can., Mr. Adams[4] has obtained the first melilite rock described from North America. It occurs as a dyke in Potsdam conglomerate. The rock, which has a fine-grained, dark groundmass, often contains phenocrysts of green and red olivine, biotite and pyroxene. The matrix in which these lie consists of small biotites, olivines and pyroxenes, between which lies a still finer aggregate of melilite, pyroxene needles and a small quantity of a colorless mineral that may be nepheline. Perofskite, apatite and magnetite are also present in it. The brown biotite is an anomite, with a small biaxial angle. It consists of an interior inclusion-free nucleus, usually with a rounded outline, surrounded by a zone filled with augite microlites and bounded by crystal faces. The olivine contains but little iron (12.65%). The red color of some grains is due to inclusions of iron oxide. Its alteration is sometimes into serpentine, but more frequently into ferriferous magnesite and breunerite, whose composition is Mg CO_3 $= 64.83$; Fe $CO_3 = 26.16$; Ca $CO_3 = 1.66$; impurities $= 7.35$. The alteration begins along cleavage cracks and proceeds inward from the peripheries of the olivine grains. The pyroxene phenocrysts are colorless and have an extinction of 42°. Like the biotite the augite grains are also bordered by a zone of a light brown color, which is of the same substance as that of the smaller phenocrysts and of the needles in the groundmass. The extinction in the zone is often 16° greater than that of the nucleus. The characteristic mineral of the rock, the melilite, possesses the peg structure and all the other peculiarities of this component of the Alnö specimens. Basal sections have rectangular or octagonal outlines, while prismatic sections are often flattened parallel to oP. The rock differs from the type alnoite in possessing no feldspar. The author thinks it is connected in some way with the Montreal volcanic center, forming Mt. Royal, which, as is well known, consists largely of eleolite syenites and related rocks. The composition of the rock follows :

SiO_2	TiO_2	Al_2O_3	Fe_2O_3	FeO	CaO	MgO	K_2O	Na_2O	CO_2	H_2O
35.91	.23	11.51	2.35	5.38	13.57	17.54	2.87	1.75		9.40

The Sanidinite Bombs of Menet and Monac.—The sanidinite bombs included in the trachytes of Menet, Cantal and of Monac,

[4] Amer. Jour. Sci., April, 1892, p. 269.

Haute-Loire, France, contain many interesting minerals, short descriptions of which are given by Lacroix.[5] In those from the first-named locality are vitreous orthoclase, anorthite microperthite, zircon in brilliant, transparent, wine-red and in colorless or brown and light rose crystals, sphene, apatite, corundum and biotite. The last named mineral is an original component of the rock yielding the bomb, while the zircon, sphene and apatite are certainly new products. The Monac sanidinites differ from those of Menet principally in being saturated with secondary substances.

Igneous Rocks from Montana.—Among the rocks found in the mountains of Montana and described by Lindgren[6] are dacite, trachytes, basalts and augite-syenites. One variety of basalt consists of fresh olivine, augite and analcite in a groundmass composed of magnetite, apatite andanalcites of a second generation. The rock was described in one of the Tenth Census Reports, where the analcite was stated to be in all probability an alteration product of nosean. The author now regards the mineral as unquestionably original.

Petrographical News.—Mr. Cole[7] describes a section of devitrified perlitic obsidian from Rocche Rosse, Lipari, in which the rock is much shattered. Around the fragments of glass thus formed spherulitic substance has resulted from the devitrification of their material. Beginning at the cracks separating the fragments the devitrification has progressed inward until a spherulitic zone now surrounds each piece of glass.——The mica schist[8] around the granite of the Schneekoppe in the Riesengebirge, Silesia, has been changed by the eruptive from a muscovite-garnet-quartz-schist to a schistose aggregate of quartz, muscovite, biotite, andalusite, and new, blood-red garnets. The biotite is in isolated small plates that are quite different in character from the flakes of muscovite adhering to the quartz grains in the original rock.——A few augite, saussurite and quartz diorites, a gabbro and several porphyries, porphyrites and diabases from the hills surrounding the Muir Glacier, in Alaska, are briefly described by Williams[9] in an appendix to Reid's account of the glacier.——In the olivine diabase of a dyke cutting the Sioux quartzite, in Minnehaha Co.,

[5]Bull. Soc. Min. d. Fr., xiv, 1892, p. 314.
[6]Proc. Cal. Acad. Sci., 2, vol. iii, p. 39.
[7]Mineralogical Magazine, ix, p. 272.
[8]Zeits. d. Deutsch. Geol. Ges. xliii, 1891, p. 780.
[9]Nat. Geog. Mag., Washington, iv, p. 63.

S. Dak., Messrs. Culver[10] and Hobbs find that the reddish to yellowish-brown diallage is strongly pleochroic.

Mineralogical News.—*General Crystallographic.*—Wyrouboff, in the continuation of his crystallographic study of closely related double salts of sulphuric, selenic and chromic acids, has reached some exceedingly interesting results bearing upon isomorphism. After carefully measuring the crystals of ten of these compounds and comparing their optical properties, he finds that while several of them are crystallographically similar these same compounds possess quite different optical characteristics. Fe $K_2(SO_4)_2$, Mn $K_2(SO_4)_2$ and Mn $K_2(SeO_4)_2$ are optically as well as morphologically similar. Since the optical properties of crystals change when they are subjected to changes of temperature it follows that these properties are dependent upon the arrangement of the molecules—upon the character of the crystal network. Isomorphous bodies are those that possess identical networks, consequently isomorphous bodies are those that are similar morphologically and at the same time optically, and in which the changes suffered under similarly changed conditions are similar. The magnesian sulphates with seven molecules of water are good examples of a truly isomorphous group. There is another kind of isomorphism embracing those bodies in which the morphological properties are similar but the optical ones different. In such bodies, since the arrangements of the molecules in the two intermingled substances are different, there should be evidence in these of optical anomalies, which are not apparent in the simple compounds, and this is found frequently to be the case. A further conclusion drawn by the author from his experiments is to the effect that while in general, substances whose chemical composition is analogous have similar crystalline forms, it does not necessarily follow that isomorphous bodies possess analogous compositions; they need merely to be built upon the same plan, possess identically arranged networks. Many of the views put forth in the paper are novel, and some of them are rather startling. We shall look forward with much interest to their discussion by German mineralogists.——The relation between symmetry and the chemical composition of crystals continues to attract the attention of mineralogists theoretically inclined. Fock[11] now suggests that the method by which the problem is to be attacked is through the aid of stereochemistry. He assumes that the crystal particles have the same symmetry as the crystal individual, and seeks

[10]Trans. Wis. Acad. Sci., viii, 1891, p. 206.

[11]Zeits. f. Kryst., xx, p. 76.

50

to trace the symmetry of the particle to the symmetry of the chemical molecule as its source. According to the conception of most chemists the carbon atom may be represented by a point with four bars extending toward the four corners of a circumscribed tetrahedron. The symmetry of the carbon molecule is thus comparable with the symmetry of the crystallized carbon—diamond. With this suggestion as a basis the author shows how the crystallization of graphite and of some of the carbonates may be explained, but at the same time he confesses that few practical results can follow from the suggestion until we know more about the composition of solid substances.

Notes.—An attempt to discover the reason for the variation in the pyramidal angles of *arsenopyrite* and to settle its composition has been made by Weibull,[12] who has examined crystals from Silfberg, Delane, and other localities in Sweden, and from the well-known occurrences in Europe. Among the Silfberg crystals three types were recognized, on the first of which the predominant forms are ∞ P and P$\widetilde{\infty}$. Their axial ratio is .6841 : 1 : 1.1910, and composition (Fe Co Ni) (S As)$_2$. On the second type the same forms are observed with the addition of $\frac{1}{2}$P$\widetilde{\infty}$, but the crystals are usually prismatic parallel to *a*. Their axial ratio is .6830 : 1 : 1.1923 and composition Fe S As. Crystals of the third type are long prismatic in the direction of *c* and are bounded by the same planes as are found in the second type. Their axial ratio is .6724 : 1 : 1.1896. Crystals from other localities show differences in composition and in axial ratio, and these differences are expressed in differences in habit. The formula best representing the composition of the mineral is thought to be Fe(S As)$_2$, and variations from it are thought to be due to inclusions in the material analyzed. If Fe(As S)$_2$ be considered the normal arsenopyrite ten per cent. of Fe S$_2$ may be replaced by Fe As$_2$, or the reverse, and the replacement will affect the axial ratio to a noticeable extent, an increase in Fe S$_2$ tending to increase the lengths of *a* and *c*. The substitution of Co and Ni for Fe affects the axes in the same way.——In an exhaustive article on the mineral deposits of Leogang in Salzburg, Buckrucker[13] gives a brief account of the region and a detailed description of the many minerals occurring therein. Thirty-two distinct species are referred to in the article, some briefly, others very extensively. Among the latter are *dolomite, aragonite, strontianite* and *celestite.* 2 E $_{Na}$ for aragonite is 30° 43.5'. 2 V $_{Na}$ for strontianite is 6° 59' 12"

[12]Ib., p. 1.

[13]Zeits. f. Kryst. xix, p. 113.

MINERALOGY AND PETROGRAPHY.[1]

The Basalt of Stempel.—Bauer's[2] description of the basalt of
Stempel, near Marburg, and its concretions and inclusions is one of
the most excellent pieces of petrographical work that has appeared in
a long time. A favorable opportunity has enabled the author to secure
a splendid suite of specimens of this rock so noted for its beautiful
zeolites. It consists of the usual constituents of basalt, viz.: plagioclase, augite and olivine in a groundmass of augite and feldspar
microlites in a base of glass. The plagioclase is andesine without
peculiar characteristics. The augite is also without special features
except that it is frequently zonally developed, with a dark-green kernel and brown-colored coats, in which the extinction decreases from 48°
to 36°. The olivine is so well bounded by crystal planes that the relations of the shapes of the cross-sections to the crystallographic axes
have been well worked out. Twins parallel to $P\infty$ are not uncommon.
The liquid inclusions, upon careful study, are found to differ from
those of the olivine of the concretions (Knollen), and the glass inclusions are learned to have a different composition from the glass forming the groundmass of the rock. One of the most interesting features
of the rock is the occurrence of amygdaloidal cavities, coated within
by a layer of glass, whose limits are sharply defined. Sometimes a
partition of this glass divides a cavity into two, and occasionally several concentric partitions give rise to a series of chambers that are
strikingly like the chambers in Idding's lithophysae. The olivine
bombs included in the rock consist largely of bronzite and chrome-
diopside grains cemented by olivine substance. The bronzite is present in two varieties, one an almost opaque greenish-brown kind, and
the other a transparent olive-green variety. Picotite is also present
quite abundantly in grains and aggregates of grains in most of the
bombs. The effect of the action of the rock magma upon its inclusions is seen in the granulation of the pyroxenes, and the effect of the
material of the bombs upon the magma is shown in the presence of microlites of hypersthene in the veins of the rock that ramify the bombs.
Since the minerals of the bombs contain characteristic inclusions not
common to lherzolitic rocks, and since, moreover, the olivine and bronzite are sometimes found in forms never seen in lherzolite, the author

[1]Edited by Dr. W. S. Bayley, Colby University, Waterville, Me.
[2]Neues Jahrb. f. Min., etc., 1891, ii, p. 156.

MINERALOGY AND PETROGRAPHY.[1]

The Basalt of Stempel.—Bauer's[2] description of the basalt of
Stempel, near Marburg, and its concretions and inclusions is one of
the most excellent pieces of petrographical work that has appeared in
a long time. A favorable opportunity has enabled the author to secure
a splendid suite of specimens of this rock so noted for its beautiful
zeolites. It consists of the usual constituents of basalt, viz.: plagioclase, augite and olivine in a groundmass of augite and feldspar
microlites in a base of glass. The plagioclase is andesine without
peculiar characteristics. The augite is also without special features
except that it is frequently zonally developed, with a dark-green kernel and brown-colored coats, in which the extinction decreases from 48°
to 36°. The olivine is so well bounded by crystal planes that the relations of the shapes of the cross-sections to the crystallographic axes
have been well worked out. Twins parallel to $P\widetilde{\infty}$ are not uncommon.
The liquid inclusions, upon careful study, are found to differ from
those of the olivine of the concretions (Knollen), and the glass inclusions are learned to have a different composition from the glass forming the groundmass of the rock. One of the most interesting features
of the rock is the occurrence of amygdaloidal cavities, coated within
by a layer of glass, whose limits are sharply defined. Sometimes a
partition of this glass divides a cavity into two, and occasionally several concentric partitions give rise to a series of chambers that are
strikingly like the chambers in Idding's lithophysae. The olivine
bombs included in the rock consist largely of bronzite and chrome-
diopside grains cemented by olivine substance. The bronzite is present in two varieties, one an almost opaque greenish-brown kind, and
the other a transparent olive-green variety. Picotite is also present
quite abundantly in grains and aggregates of grains in most of the
bombs. The effect of the action of the rock magma upon its inclusions is seen in the granulation of the pyroxenes, and the effect of the
material of the bombs upon the magma is shown in the presence of microlites of hypersthene in the veins of the rock that ramify the bombs.
Since the minerals of the bombs contain characteristic inclusions not
common to lherzolitic rocks, and since, moreover, the olivine and bronzite are sometimes found in forms never seen in lherzolite, the author

[1] Edited by Dr. W. S. Bayley, Colby University, Waterville, Me.
[2] Neues Jahrb. f. Min., etc., 1891, ii, p. 156.

MINERALOGY AND PETROGRAPHY.[1]

The Basalt of Stempel.—Bauer's[2] description of the basalt of Stempel, near Marburg, and its concretions and inclusions is one of the most excellent pieces of petrographical work that has appeared in a long time. A favorable opportunity has enabled the author to secure a splendid suite of specimens of this rock so noted for its beautiful zeolites. It consists of the usual constituents of basalt, viz.: plagioclase, augite and olivine in a groundmass of augite and feldspar microlites in a base of glass. The plagioclase is andesine without peculiar characteristics. The augite is also without special features except that it is frequently zonally developed, with a dark-green kernel and brown-colored coats, in which the extinction decreases from 48° to 36°. The olivine is so well bounded by crystal planes that the relations of the shapes of the cross-sections to the crystallographic axes have been well worked out. Twins parallel to $P\widetilde{\infty}$ are not uncommon. The liquid inclusions, upon careful study, are found to differ from those of the olivine of the concretions (Knollen), and the glass inclusions are learned to have a different composition from the glass forming the groundmass of the rock. One of the most interesting features of the rock is the occurrence of amygdaloidal cavities, coated within by a layer of glass, whose limits are sharply defined. Sometimes a partition of this glass divides a cavity into two, and occasionally several concentric partitions give rise to a series of chambers that are strikingly like the chambers in Idding's lithophysae. The olivine bombs included in the rock consist largely of bronzite and chrome-diopside grains cemented by olivine substance. The bronzite is present in two varieties, one an almost opaque greenish-brown kind, and the other a transparent olive-green variety. Picotite is also present quite abundantly in grains and aggregates of grains in most of the bombs. The effect of the action of the rock magma upon its inclusions is seen in the granulation of the pyroxenes, and the effect of the material of the bombs upon the magma is shown in the presence of microlites of hypersthene in the veins of the rock that ramify the bombs. Since the minerals of the bombs contain characteristic inclusions not common to lherzolitic rocks, and since, moreover, the olivine and bronzite are sometimes found in forms never seen in lherzolite, the author

[1] Edited by Dr. W. S. Bayley, Colby University, Waterville, Me.
[2] Neues Jahrb. f. Min., etc., 1891, ii, p. 156.

MINERALOGY AND PETROGRAPHY.[1]

The Basalt of Stempel.—Bauer's[2] description of the basalt of Stempel, near Marburg, and its concretions and inclusions is one of the most excellent pieces of petrographical work that has appeared in a long time. A favorable opportunity has enabled the author to secure a splendid suite of specimens of this rock so noted for its beautiful zeolites. It consists of the usual constituents of basalt, viz.: plagioclase, augite and olivine in a groundmass of augite and feldspar microlites in a base of glass. The plagioclase is andesine without peculiar characteristics. The augite is also without special features except that it is frequently zonally developed, with a dark-green kernel and brown-colored coats, in which the extinction decreases from 48° to 36°. The olivine is so well bounded by crystal planes that the relations of the shapes of the cross-sections to the crystallographic axes have been well worked out. Twins parallel to $P\widetilde{\infty}$ are not uncommon. The liquid inclusions, upon careful study, are found to differ from those of the olivine of the concretions (Knollen), and the glass inclusions are learned to have a different composition from the glass forming the groundmass of the rock. One of the most interesting features of the rock is the occurrence of amygdaloidal cavities, coated within by a layer of glass, whose limits are sharply defined. Sometimes a partition of this glass divides a cavity into two, and occasionally several concentric partitions give rise to a series of chambers that are strikingly like the chambers in Idding's lithophysae. The olivine bombs included in the rock consist largely of bronzite and chrome-diopside grains cemented by olivine substance. The bronzite is present in two varieties, one an almost opaque greenish-brown kind, and the other a transparent olive-green variety. Picotite is also present quite abundantly in grains and aggregates of grains in most of the bombs. The effect of the action of the rock magma upon its inclusions is seen in the granulation of the pyroxenes, and the effect of the material of the bombs upon the magma is shown in the presence of microlites of hypersthene in the veins of the rock that ramify the bombs. Since the minerals of the bombs contain characteristic inclusions not common to lherzolitic rocks, and since, moreover, the olivine and bronzite are sometimes found in forms never seen in lherzolite, the author

[1]Edited by Dr. W. S. Bayley, Colby University, Waterville, Me.
[2]Neues Jahrb. f. Min., etc., 1891, ii, p. 156.

metacinnabarite from New Almaden, Cal., have given Melville[19] an opportunity to measure and to analyze its crystals. The mineral occurs in steep rhombohedral forms attached to quartz crystals, which in turn coat cinnabar crystals, resting in a compact mass of this substance and quartz. The terminations of the crystals are differently modified, the analogue pole containing principally the basal plane and rhombohedra, and the antilogue·pole mainly steep scalenohedra. The analysis, made on impure material, gave:

S	Hg	Fe	Co	Zn	Mn	CaCo$_3$	Res.	Org. mat.	Total
13.68	78.01	.61	tr.	.90	.15	.71	4.57	.63	= 99.26 ·

The organic matter was in the form of little black spheres imbedded within the crystals.——If the orthopinacoids of the members of the *heulandite* group be made the orthodome $\frac{1}{2}P\infty$, Rinne[20] shows that its members may be regarded as forming an isomorphous group with stilbite, harmotome and phillipsite. The axial ratios of the various minerals come into accord with this view if half of *c* is taken as the unit in the members of the stilbite sub-group. The chemical composition of the different minerals is also similar enough to oppose no objection to the idea, the stilbites being mixtures of $R'' Al_2 Si_6 O_{16} + 6Aq$ and $R_2 Al_2 Al_2 Si_4 O_{16} + 6Aq$, and the heulandites $R Al_2 Si_6 O_{16} + 6Aq$ for heulandite proper, and $R Al_2 Si_6 O_{16} + 5\frac{1}{2}Aq.$ for epistilbite and brewsterite. The physical properties of all the substances mentioned are quite alike and their optical peculiarities are not different.——The *chloritoid* of a graywacke schist from the Champion Mine, Mich., is similar in many respects to masonite, according to Keller and Lane.[21] It is undoubtedly triclinic with *B* inclined 20° to the basal cleavage. Its pleochroism is $C =$ yellow, $B =$ blue, $A =$ green. An analysis gave:

SiO$_2$	TiO$_2$	Al$_2$O$_3$	Fe$_2$O$_3$	FeO	MnO	MgO	CaO	K$_2$O	Na$_2$O	H$_2$O	Total
24.29	.28	34.00	10.55	20.52	tr.	1.29	.59	.97	.35	6.75	=99.59

Its hardness is 6.5 and density = 3.552.——Streng[22] again attempts to solve the composition of *melanophlogite* and succeeds in showing that the sulphur in its material is not in the form of sulphate but is more prob-

[19]Bull. U. S. Geol. Survey, No. 78, p. 80.

[20]Neues Jahrb. f. Min., etc., 1892, i, p. 12.

[21]Amer. Jour. Sci., Dec., 1891, p. 499.

[22]Neues Jahrb. f. Min., etc., 1891, ii, p. 211.

ably present as SiS_2, combined in some way with SiO_2, in the proportion $SiS_2 + 40\ SiO_2$. G. Friedel,[23] on the other hand, insists upon regarding the sulphur as occurring in the form of sulphate.——The *sigterite* from Sigterö, Norway, described by Rammelsberg[24] a short time ago as a new mineral, is acknowledged by this savant and by Tenne to be a mixture of eleolite and albite.——A greenish-white fibrous *talc* from Madagascar[25] has the composition $SiO_2 = 62.3$; $FeO = 2.6$; $MgO = 29.4$; $H_2O = 5.1$.——On a specimen of *dioptase*[26] in calcite from Central Africa Jannetaz has recognized octahedra of silver. This is the first report of the existence of native silver in that quarter of the globe.——Crystals of *barite* from Smithton and Sedalia, Pittis Co., Mo., consist of colorless portions enclosing yellow or white bands, in the latter of which Luedeking[27] and Wheeler find a large quantity of strontium and a small amount of ammonium sulphate. The composition of the crystals is $Ba\ SO_4 = 87.2$; $Sr_2SO_4 = 10.9$; $Ca\ SO_4 = .2$; $NH_4So_4 = .2$; $H_2O = 2.4$.——In consideration of the importance given by Tschermak to *meionite* in his discussion of the scapolite group Kenngott recalculates the formula of the mineral from new analyses recently published and derives $Ca_7\ Al_{10}\ O_{23}\ Si_{11}\ O_{27}$. He evidently places but little confidence in the Tschermak theory.——By mingling solutions of chromates, tungstates, molybdates, sulphates and selenates and studying the mixed crystals resulting Retgers[28] has shown that their alkaline and other salts are isomorphous, and that consequently when they are found as minerals they should all be placed in one group, which is trimorphous. The tellurates, on the other hand, are not isomorphous with any of the above mentioned compounds.—— The walls of cavities of the leucite basalt from the south side of Lake Laach are covered with brilliant little crystals that have been carefully examined by Busz.[29] They are *hematites* on which are implanted *rutile* crystals and little colorless *olivines* with a tabular habit parallel to $\infty P\overline{\infty}$. All are supposed to be products of sublimation.——A. Schmidt[30] records the results of observations on pebbles of *zircon*, *almandine* and epidote from Adelaide, Australia. The zircon has a

[23]Bull. Soc. Fr. d. Min., xiv, p. 74.

[24]AMERICAN NATURALIST, 1890, p. 1189.

[25]Jannetaz, Bull. Soc. Fr. d. Min., xiv, p. 66.

[26]Ib., xiv, p. 67.

[27]Amer. Jour. Sci., Dec., 1891, p. 495.

[28]Neues Jahrb. f. Min., etc., 1892, i, p. 56.

[29]Zeits. f. Kryst., 1891, xix, p. 24.

[30]Ib., 1891, xix, p. 56.

density of 4.695 and a composition of $ZrO = 67.31$; $SiO_2 = 33.42$. The author also describes cubical and octahedral crystals of pyrite from Porkura, Hungary.——*Enargite* from Cerro Blanco, Atacama, Chile, has a density of 4.51. It contains $S = 32.21$; $As = 18.16$; $Cu = 47.96$; $Fe = 1.22$; $Zn = .57$.——The *amber-like* substance[21] occurring in the sands of Cedar Lake, near the mouth of the Saskatchewan River, in Canada, has been found by Harrington[22] to have the following composition: $C = 79.96$; $H = 10.46$; $O = 9.49$; $As = .09$. Its hardness is 2.5 and density 1.055. From its reaction with solvents the author concludes it to be *retinite.*——Lacroix and Baret[23] find *bertrandite* at Mercerie in the Commune of La-Chapellesur-Erdre, France. It occurs in crystals elongated parallel to the base, associated with orthoclase, albite, quartz and apatite, in a granitite.

[21] Neufville, Ib., 1891, p. 75.

[22] Amer. Jour. Sci., Oct., 1891, p. 332.

[23] Bull. Soc. Franç. d. Min., xiv, p. 189.

September 1st, 1892.

MINERALOGY AND PETROGRAPHY.[1]

Thermometamorphism.—A very brief but extremely interesting review of the effects of thermometamorphism in the acid and basic lavas in contact with granite and granophyre in the Lake District, Eng., is given by Harker.[2] The altered zone surrounding the latter rocks is about ½ of a mile wide. On its outer periphery only the secondary constituents of the basic lavas have been affected by the contact action. This leads the author to the statement that the substances most susceptible to thermal agency are those formed under ordinary meteoric conditions, minerals of direct igneous origin being more refractory. In the rocks under discussion but little change in chemical composition has taken place as a consequence of their metamorphism, except that there is a loss of water and a gain in boron quite near the contact. The mineralogical changes noted are the production of biotite from the chloritic decomposition products of pyroxene and the formation of clear feldspar from the original turbid mineral. In addition to these, quartz, green hornblende, actinolite, tremolite, augite, sphene, rutile, magnetite and pyrite have also resulted from the metamorphic processes. The characteristic contact minerals of sedimentary rocks, cyanite, andalusite and garnet are practically absent. Their abundant presence in sedimentary rocks is thought to be due to the fact that these contain but a small proportion of alkalies, as a result of the long continued chemical degradation to which they have been subjected, and that since feldspar, which the author regards as a characteristic contact mineral, could not form, only aluminous new products are possible. The careful study of the altered rocks indicates that there was very little interchange of substance between different portions of the original rock, except between those parts that were immediately adjacent. The preservation of minute structures, such as fluxion lines and spherulitic aggregates, point to this conclusion. In the case of the acid lavas the material very near the metamorphosing mass consists principally of a fine-grained aggregate of feldspar and quartz. Aluminous and ferruginous compounds were absent from the original lava; they are likewise absent from its altered phases.

[1]Edited by Dr. W. S. Bayley, Colby University, Waterville, Me.
[2]Bull. Geol. Soc. Amer., Vol. iii, 1892, p. 16.

Cellular Epidote in Granite.—Three granite sections from Wrangell Island, Alaska, the Pelly River in the Yukon District, and from the Stikine River, in the Coast Range of British Columbia, are described by Adams[3] as exhibiting several interesting peculiarities. All possess large quantities of epidote, which has a rudely outlined crystal form. In the specimen from the first mentioned locality the epidote is intergrown with allanite, with the latter on the interior. The most peculiar feature in connection with the epidote is its cellular structure, it being merely a skeleton of this substance enclosing small, elbow-shaped areas of quartz, feldspar, etc. Calcite and muscovite grains were noted with the same structure. In the case of the mica the inclusions occupy such a large portion of the area enclosed within the outline of the mineral that the muscovite appears as an assemblage of detached fragments, optically continuous with one another. Upon discussing the origin of the cellular minerals the author is compelled to the conclusion that they have all been formed since the consolidation of the rocks in which they occur. Since these all show evidence of having undergone slight crushing, it may be that the growth of the minerals is dependent somehow upon the reactions set up during the crushing. As all the constituents in these granites are fresh, the conclusion that the calcite, muscovite and epidote are secondary is an interesting one.

Petrographical Notes.—In an article descriptive of Chilean ore deposits Möricke[4] gives a few petrographical notes on a hornblende-biotite granite, a tourmaline granite, a quartz diorite, a quartz porphyry, and two other rocks of special interest. One is a quartz tourmaline rock in which a sort of groundmass of the former mineral encloses small idiomorphic crystals of the tourmaline. It is presumably an eruptive. The other is a perlitic pitchstone from Guanaco, with large phenocrysts of plagioclase and sanidine and a few flakes of biotite. Its unique feature is the possession of gold in skeleton crystals scattered through the glassy matrix, enclosed in the spherulites and included in the fresh feldspar.——Masses of an azure blue saccharoidal[5] rock occur imbedded in a granular serpentine at a point on the Gila River, 40 miles west of Silver City, N. M. In the thin section these masses are found to be composed of a granular, colorless

[3]Canadian Rec. of Sci., Sept., 1891, p. 844.
[4]Min. u. Petrog., Mitth. xii, p. 195.
[5]Merrill and Packard. Amer. Jour. Sci., April, 1892, p. 279.

pyroxene intermingled with calcite, with the pyroxene more or less altered into serpentine. A fragment of the rock free from calcite and serpentine gave: $SiO_2 = 54.30$; $MgO = 18.33$; $FeO = 1.11$; $CaO = 25.00$, a composition corresponding to $Ca\,Mg\,(SiO_3)_2$. The blue color is supposed to be due to the ferrous iron present in the pyroxene.

Two New Rocks.—Boninite is a bronzite limburgite from Peel Island, one of the Bonine group, near Japan. It is described by Petersen[6] and Kikuchi[7] as consisting of phenocrysts of olivine, bronzite and a few augites imbedded in a glass full of crystallites, some of which are sanidine. The rock is closely related to sanukite.[8] Mijakite from Mijakeshima is an andesite with a reddish brown pyroxene, supposed to be triclinic, feldspar and glass, forming a groundmass in which are porphyritic crystals of bytownite, a little augite, hypersthene and biotite. The composition is:

SiO$_2$	Al$_2$O$_3$	Fe$_2$O$_3$	FeO	MnO	MgO	CaO	Na$_2$O	K$_2$O	Loss	Total.
50.87	21.98	5.85	5.09	1.45	1.38	9.12	2.85	22	.43	= 99.24

Optical Anomalies.—In a prize volume[9] issued by the Fürstlich Jablonowski Society of Leipzig, Brauns discusses critically and in great detail the various theories proposed in explanation of optical anomalies and gives a resumè of all the work done on individual minerals exhibiting the phenomena. About seventy substances in which anomalies have been discovered are treated in the second part of the volume, while in the first part the space is devoted to the historical and critical discussion of the theories. The anomalous bodies are divided into five groups according as the cause of their peculiarities is differently orientated lamellæ; dimorphous enantiotropism of their substance, strain, isomorphous mixture or loss of water. In an appendix are grouped those minerals the cause of whose anomalies is unknown.——Pyrenaite, the black garnet occurring in a black limestone at the Pic d'Eres Lids, Pyrenees, show such regular anomalies that Mallard[10] is enabled to determine the optical constants of the sub-

[6]Jahrb. Hamburg. wissensch. Anst. viii, 1891, p. 1.

[7]Jour. Coll. of Sci., Imp. Univ. Japan, iii, 1889, p. 67. Ref. Neues Jahrb. f. Min., etc., 1892, i, pp. 311 and 313.

[8]AMER. NATURALIST, 1891, p. 868.

[9]Die Optischen Anomalien der Krystalle. Leipzig, S. Hirzel, 1891. Pl. 6, pp. 10 and 370.

[10]Bull. Soc. Franç. d. Min., xiv, p. 293.

stance whose six orthorhombic pyramids build up the perfect dodecahedral crystals. Its mean index of refraction is 1.74 and its optical
angle 2 V = 56° 5′. The author regards his observations as settling
the question as to the cause of optical anomalies in garnet in favor of
his own theory and in opposition to the theories of Klein and of
Brauns, the former of whom regards them as due to the dimorphism
of the garnet molecule and the tension resulting from its attempt to
pass to a more stable form than that in which it crystallized, and the
latter as due to isomorphous mixtures.

Examination of thin sections of beryl crystals from the Ilmengebirge in the Urals leads Karnojitzky[11] to the belief that the anomalies
discoverable therein are dependent directly upon the limiting faces of
the crystals. When these differ the character of the internal structure differs, as is also often true in the case of garnet. An optically
anomalous beryl crystal consists of several elemental individuals, the
number and position of which correspond closely with the number
and positions of the limiting planes of the crystals. The positions of
the individuals preclude the notion of twinning. The author thinks
the anomalies due to the isomorphous mixture of the beryl substance
with some other, probably tourmaline.——In a section of dioptase cut
parallel to the base the same investigator[12] found uniaxial areas distributed among the normal biaxial areas in such a way as to convince him
that the interior structure of the mineral is determined to some extent
by its exterior form, as in the case of the beryl.

Mineralogical News.—Genth and Penfield[13] have obtained
hübnerite crystals from two localities near Silverton, Col., from White
Oaks, N. M., and from Nye Co., Nev. Those from the North Star
Mine, near Silverton, were doubly terminated, so that by their measurement the axial ratio, .8362 : 1 : .8668, was determined. Cleavage
sections parallel to ∞ P∞̄ extinguish at about 17° from *c* in the obtuse
β. which direction is the axis of least elasticity. Pleochroism is
marked, being green parallel to *C* and yellowish brown parallel to *B*.
Density is 6.713 and composition $WO_3 = 74.75$; $FeO = 2.91$; $MnO
= 21.93$; $CaO = .11$; $MgO = $ tr. *Bismuthite* from the phenacite
locality of Mount Antero, Col., and *hessite* from the Refugio Mine,
Jolisco, Mex., are also briefly described by the same authors. They
were, however, too impure to yield good analytical results. A *natrolite*

[11]Zeits. f. Kryst., 1891, xix, p. 209.

[12]Ib., 1891, xix, p. 593.

[13]Amer. Jour. Sci., March, 1892, p. 184.

associated with eudialite, etc., at Magnet Cove, Ark., is in large color-less crystals whose analysis gave: $SiO_2 = 47.97$; $Al_2O_3 = 26.51$; $Na_2O = 15.98$; $H_2O = 9.81$.

Yeates and Ayres[14] have come into the possession of sufficient quantities of *plattnerite* to enable them to describe it in some detail. The mineral is associated with limonite and white pyromorphite at the "You Like" lode, near Mullan, Idaho. The pyromorphite is in veins cutting nodules of the plattnerite and in little crystals imbedded in them. The color of the lead oxide is iron black. Its streak is chestnut-brown, its hardness 5.5 and density 8.56. An analysis showed the following composition: $Pb = 83.20$; $Ag = tr.$; $Cu = .14$; $Fe Al = 1.20$; $O = 12.93$; $Ins. = .82$, besides Ca and Mg. Upon breaking open some of the nodules little cavities were found in it, whose walls were covered by druses of tiny crystals. These plattnerite crystals are tetragonal and isomorphous with the members of the rutile group. $a : c = 1 : .67643$. They are prismatic with $\infty P\infty$, $3P\infty$ and sometimes oP and $\frac{1}{2}P$.

In another contribution to the discussion of the constitution of micas and chlorites Clarke and Schneider[15] communicate results of analyses of *waluewite*, *clinochlor* and *leuchtenbergite* from Slatoust, of diallage *serpentine* from Syssert, and of white *mica* from Miask, in the Urals. One of the conclusions based upon the action of the chlorites toward reagents is to the effect that their composition cannot be explained in terms of the Tschermak theory.

At Placerville, Eldorado Co., Cal., is a vein of quartz cutting quartzite. The vein is much decomposed and contains pockets of red earth in which are numbers of *quartz*[16] crystals, some of immense size. Many bear inclusions of chlorite arranged in zones marking successive stages of growth, and others contain hollows that are moulds of groups of some rhombohedral mineral, probably siderite. *Brookite* and *octahedrite* are also found in these pockets, sometimes loose in the clay, sometimes implanted upon the quartz or included within it. In the same article there is described an immense *monazite* crystal from Perm, Russia, and enormous rubies from Moguk, near Mandalay, Burma.

The *bournonite* of Nagybanya, in Hungary, is associated with zinc, lead and other sulphides, siderite and quartz. According to A. Schmidt[17] two types of the bournonite occur, the prismatic and the

[14]Ib., 1892, May, p. 407.

[15]Ib., 1892, p. 378.

[16]Ib., Feb., 1892, p. 329.

[17]Zeits. f. Kryst., xx, p. 152.

tabular, both of which are rich in forms. Two new planes $\frac{1}{2}P\overline{\infty}$ and
$2P\underset{\infty}{\smile}$ raise the number of forms known to occur in the species to 75.

In the carbonaceous mica schists near the contact with granite in
the Müglitzthal, S. E. of Dresden, Beck[18] has discovered small crystals of *brookite.* The original rock from which the schists were made
contained rutile needles. These afforded the material for the brook-
ites, which are found lying with their flat sides parallel to the cleavage
planes of the schists.

The striations parallel with and perpendicular to the octahedral
edges of the *magnetite* crystals from Mineville, N. Y., are ascribed by
Kemp[19] to etching agents. The striations parallel to the edges are
usually referred to twinning, but in the present case it is probable that
Kemp's explanation is the correct one, for when unstriated crystals
are subjected to the influence of HCl and H_2SO_4 they become covered
with striations like those occurring in nature.

A lot of very pure *cordierite* from coarse veins cutting gneiss at
Guilford, Ct., has been analyzed by Farrington,[20] who finds in it :

SiO_2	Al_2O_3	Fe_2O_3	FeO	MnO	MgO	H_2O
49.44	32.97	.35	5.11	.32	10.39	1.65

corresponding to H_2O, $4(Mg\ Fe)O$, $4\ Al_2O_3$, $10\ SiO_2$.

Treadwell[21] has made a new analysis of *milarite* that yielded him :

SiO_2	Al_2O_3	CaO	K_2O	Na_2O	H_2O	MgO
72.79	10.12	11.32	4.32	.26	1.19	tr.

This corresponds to R_2O, $2\ CaO$, Al_2O_3, $12\ SiO_2$.

Scalenohedral and prismatic crystals of *calcite* from Niederraben-
stein, in Saxony, are described by Beckenkamp.[22] The prismatic ones
are tetartohedral and hemimorphic.

Miscellaneous.—Dunnington,[23] after careful quantitative analyses of seventy-two soils from various localities as widely separated as
the Sandwich Islands and Palestine, comes to the conclusion that all

[18]Neues Jahrb. f. Min., etc., 1892, i, p. 159.
[19]Zeits. f. Kryst , xix, p. 193.
[20]Amer. Jour. Sci., Jan., 1892, p. 18.
[21]Neues Jahrb. f. Min., etc., 1892, i, p. 167.
[22]Zeits. f. Kryst., xx, p. 163.
[23]Amer. Jour. Sci., Dec., 1891, p. 491.

contain titanium, which of course must necessarily exist as widely spread in the rocks from which the soils were made.

Small crystals of *melilite* have been detected by Bödländer[24] in a mass formed by the melting of Portland cement in an oven with a lining containing 63%–88% of SiO_2. The raw materials used as the charge consisted of a mixture of magnesia, limestone and clay. The crystals were imbedded in a mass of olivine(?), magnetite, mica and apatite. The crystals were found to be optically positive and to have the composition :

SiO_2	Al_2O_3	Fe_2O_3	CaO	MgO	K_2O	Na_2O
37.96	9.46	2.93	34.75	12.77	1.53	.64

Becke[25] gives brief but very definite directions as to the use of his method of distinguishing between quartz, orthoclase and plagioclase by etching with hydrofluoric acid and staining their etched surfaces.

Schrauf[26] describes a method of combining microscope and reflection goniometer in such a way that minute crystals may easily be studied and measured.

[24]Neues Jahrb. f. Min., etc., 1892, i, p. 53.
[25]Min. u. Petrog., Mitth. xii, p. 257.
[26]Zeits. f. Kryst., xx, p. 90.

October 1st, 1892.

MINERALOGY AND PETROGRAPHY.[1]

The Geology of the Kaiserstuhlgebirge, by Gráeff,[2] contains
a resumé of the facts known concerning the structure of this celebrated
region, and a brief synopsis of the characteristics of the interesting
volcanic rocks occurring there. The tephrites, basanites, phonolites,
limburgites, nephelinites and leucites found in dykes and flows in the
mountains are described only briefly, as they are all well-known to
petrographers. The loess, tufas and the crystallized limestone, the latter
of which forms the central portion of the heights, are treated as briefly,
except that in relation to the origin of the limestone the author enters
upon a discussion to show that it is probably a metamorphosed Jurassic
rock. The most interesting portion of the paper is that which
describes the inclusions in the eruptives. These are gneiss, granite,
eleolite-syenite, and fragments of the volcanic rocks. They have all
been more or less altered by the eruptive in which they are imbedded.
The wollastonite and melanite crystals, both very common in the phonolite, are thought to be the remnants of metamorphosed limestone
fragments. The most striking inclusions are those found in a phonolite dyke near Obenbergen. They are often coarsely granular, and
sometimes have rounded outlines. Their mineral constituents are the
same as those of the including phonolite; but usually some one or
more of them is completely lacking. Orthoclase, hauyne and nepheline are the most abundant components, and hauyne the most persist-
ent, entire inclusions sometimes consisting almost wholly of large
idiomorphic hauyne crystals. Graeff supposes them to be the cooled
intratellurial portions of the magma, which on the surface yielded
phonolite, that, after solidification, were brought to the surface by a second eruption of the same material. He believes the olivine bombs in
basalts have an analogous origin, and that they are not simply concretions of the basic minerals of this rock.

A Cyanite-Garnet-Granulite from the Tirolese Alps.—
This rock, obtained some time ago by Cathrein, has been examined
microscopically by Ploner.[3] The garnet and cyanite are both in large

[1]Edited by Dr. W. S. Bayley, Colby University, Waterville, Me.
[2]Mitth. der Gross Badischen Geol., Landesanst 2, xiv, p. 405.
[3]Min. u. Petrog., Mitth. xii, p. 318.

grains, the former in dodecahedral crystals that have in many instances been shattered by pressure, and the latter in bent plates with the usual features of cyanite. Biotite encircles both of these minerals, notably the garnet, as a sort of zone. The groundmass in which these components lie is an aggregate of oligoclase, orthoclase and quartz, sometimes the monoclinic and at other times the triclinic feldspar predominating. Rutile is present in the rock as inclusions in the garnet, the cyanite and the biotite, as an alteration product of the mica, and as crystals in the quartz-feldspar aggregate. Muscovite, ilmenite, zircon and leucoxene are also present in small quantities.

Tufaceous Slates from Wales.—Among the sedimentary roofing slates of North Wales Hutchings[4] finds some that appear to be composed principally of andesitic and rhyolitic ash, consisting of fragments of lapilli, of feldspar and of sedimentary rocks imbedded in a paste of chlorite, small rods of sericite and minute grains of garnet, besides a little quartz and calcite. The most essential differences between these slates and those of sedimentary origin are with respect to their titanium constituents; in the ashes sphene and anatase being the most important, and in the true slates the so-called "slate-needles." These latter are thought by the author to occur only as decomposition products of biotite, and to be limited in their occurrence to water deposited fragmentals. The feldspar in the rocks under discussion are changed to white mica, chlorite and calcite. Secondary orthoclase and plagioclase often coat tiny cavities in the rock.

Alteration Products of Diabase from Friedensdorf.—The clefts in the diabase of Friedensdorf, near Marburg, are covered with little crystals of albite, analcite, natrolite, prehnite and calcite, all of which minerals occur also in the body of the rock. According to Brauns[5] they are decomposition products of the diabase plagioclase, and are due to the action of water containing carbon-dioxide upon this feldspar. Microscopic sections show the original plagioclase surrounded by fresh albite and filled with little nests of the other secondary substances mentioned. The process of the alteration is outlined by the author, who also shows the chemical relations existing between the new substances and the material from which they were derived. The diabase originally contained in addition to the plagioclase, both monoclinic and orthorhombic pyroxenes, olivine and titanic magnetite.

[4] Geol. Magazine, 3, ix, 1892, pp. 145–335.
[5] Neues. Jahrb. f. Min., etc., 1892, ii, p. 1.

The olivine and the orthorhombic pyroxene are serpentinized and the plagioclase altered as already indicated.

Camptonite Dykes in Maine.—In the gneiss of Androscoggin County, Maine, especially in the vicinity of Lewiston and Auburn, are a number of small dykes, some of which are of normal diabase, while others consist of camptonites. Olivine is abundant in several of the latter, and in such large grains as to be readily detected in the hand specimen. Olivine and augite are frequently in phenocrysts, while the last named mineral, hornblende and plagioclase make up the large part of the groundmass of the lamprophyres. An analysis of material from one of the dykes yielded Merrill and Packard: [6]

SiO_2	TiO_2	Al_2O_3	Fe_2O_3	FeO	MnO	CaO	MgO	K_2O	Na_2O	P_2O_5	CO_2	H_2O
39.32	1.70	14.48	2.01	8.73	.71	8.30	11.11	.87	3.76	.61	5.25	2.57

Predazzites and Pencatites.—Twenty specimens of predazzites and pencatites from various localities have been examined by Lenecek[7] in order to determine whether the rocks contain brucite or not. The sections of the true predazzites were found to have a calcite groundmass, scattered through which are fibres of hydromagnesite, supposed to be pseudomorphs after periclase, since cross sections of groups of fibres have a regular outline, and since one section of pencatite from Canzacoli shows periclase crystals more or less changed to serpentine. The dark pencatites differ from the predazzites in containing a large quantity of marcasite, to whose opacity the dark color of the rock is due. Besides the constituents already mentioned there are in both rocks many small grains of colorless silicates that may be pyroxenes, amphibole and olivine. Serpentine veins also cut both rocks, and brucite plates are not uncommon as the lining of little cracks.

Petrographical News.—Around the granite boss of Cima d'Asta, as around the other eruptive masses of eastern South Tyrol, there are abundant evidences of contact action in the contiguous sedimentaries, the contact rocks being not different in their essential characteristics from those surrounding the Adamello tonalite. The tonalite gneiss of the Adamello region is a pressure gneiss, occurring along lines, which were the slipping directions in the eruptive.[8]

[6] Am. Geol., x, 1892, p. 49.
[7] Min. u. Petrog. Mitth., xii, p. 429.
[8] Salomon. Min. u. Petrog., Mitth. xii, p. 408.

At last Rosenbusch[9] has replied to Michel Levy's criticism of his classification of massive rocks. In this reply the author first corrects some misstatements made in Levy's brochure, and then discusses the questions of priority which the French savant raises. After effectually disposing of these points Rosenbusch gives the reasons that led him to suggest the separation of massive rocks into the three classes, the plutonic, the volcanic, and the dyke rocks, and states that the recent work of all petrographers has strengthened his determination to hold to this classification.

The granite, porphyry, schist and clastic rock boulders occurring in the various conglomerates aud breccias of the "Flysch" in Switzerland have been thoroughly studied by Sarasin,[10] who recognizes among them many that are identical in substance with rocks in the Southern Alps. This fact leads him to suggest that the middle Alps were not elevated to anything like their present height at the time when the conglomerates and breccias were formed, but that there was then an unimpeded course from the Southern Alps to the northern side of the Northern Alpine ranges.

In an article entitled The Geology of the Massive Rocks of the Island of Cyprus, Bergeat[11] describes with very little detail diabase, gabbro, wehrlite, serpentine, andesite, liparite, trachyte, and andesitic and lipariatic tufas, all of which occur in some quantity on the Island. All are very much altered.

In a block that fell from the walls of the Legbachthal, Oberpinzgau, in the central Alps, Weinschenk[12] found a small dyke of much altered kersantite. On the contact of the dyke with the intruded biotite feldspar schist the latter is changed to an aggregate of epidote, quartz, feldspar and muscovite.

Hibsch[13] describes from Southern Paraguay a sandstone, a quartz porphyry and a nepheline-basalt.

Josephinite, a New Nickel-Iron Alloy.—*Josephinite*[14] occurs as magnetic pebbles in the placer gravel of a stream in Josephine and Jackson Counties, Oregon. The pebbles consist of a greenish-black siliceous substance intermingled with grayish-white areas of the alloy.

[9] Ib., xii, p. 351.
[10] Neues. Jahrb. f. Min., etc., B. B. viii, p. 180.
[11] Min. u. Petrog., Mitth. xii, p. 263.
[12] Min. u. Petrog., Mitth. xii, p. 328.
[13] Ib., xii, p. 253.
[14] Amer. Jour. Sci., June, 1892, p. 509.

The siliceous matter is partly serpentine and partly a silicate, insoluble in acid, possibly an impure bronzite. The alloy has a composition corresponding to $Fe_2 Ni_5$. Chromite, magnetite and troilite are also present in the pebbles, the first two as granules in the silicates. The alloy is gray, malleable and sectile, and has a hardness of 5. Its origin is probably terrestrial.

Crystallography.—On crystals of *vesuvianite* from the blocks of Monte Somma, Boecker[15] finds seven new forms and recognizes a tabular type hitherto undescribed. The new forms detected are $\frac{1}{4}P\infty$, $\frac{8}{3}P$, $\frac{8}{3}P$, $\frac{7}{14}P$, $\frac{5}{18}P\frac{5}{18}$, $\frac{9}{7}P\frac{9}{7}$, and $\frac{11}{4}P\frac{11}{4}$. He describes also transparent green crystals of the same substance implanted in granular yellowish-green vesuvianite from Lermatt.

On *topas* from near Miass in the Ilmen Mountains, S. Urals, Souheur[16] reports a large number of new planes in the prismatic and the pyramidal zones, and that between $P\bar{\infty}$ and $\frac{1}{2}P$. The crystals are from Redikorzew's topaz mine, where they are associated with ilmenorutile, black tourmaline, and muscovite on an amazonite-bearing granite.

The plane $P\breve{2}$ has been discovered by Pelikan[17] in *sulphur* crystals, implanted on antimonite from Allchar, Macedonia. Measurements of cleavage pieces of meteoric iron incline Linck[18] to the belief that the twinning of the iron is parallel to the plane 2O2.

Mineralogical Notes.—Another calculation of the formula of *tourmaline* from published analyses leads to the suggestion by Kenngott[18] that the various members of the tourmaline group are isomorphous mixtures of the compounds $3R_2O.\ SiO_2 + 5\ (R_2O_3.\ SiO_2)$ and $2\ (3RO.\ SiO_2) + R_2O_3.\ SiO_2$. The red tourmaline from Rumford, Me., may be regarded as the first end member of the series. The last end member is not yet known.

New analyses of *pseudobrookite*[19] from the Siebenbürgen yield no magnesia. Crystals from this locality, like those from Norway, thus consist simply of iron and titanium oxides. They are tabular with $\infty P\bar{\infty}$, $\infty P\breve{\infty}$, $\infty P\breve{2}$, ∞P, $P\bar{\infty}$, $\frac{1}{3}P\breve{\infty}$ and $\frac{1}{2}P$, of which the latter is

[15]Zeits. f. Kryst., xx, p. 225.

[16]Zeits. f. Kryst., xx, 1892, p. 232.

[17]Min. u. Petrog., Mitth., xii, p. 344.

[17]Zeits. f. Kryst., xx, p. 209.

[18]Neues. Jahrb. f. Min., etc., 1892, ii, p. 44.

[19]Traube. Zeits. f. Kryst., xx, 1892, p. 327.

new. Their axial ratio is .98123 : 1 : 1.12679. The mineral is found in clefts of an andesite, or in the rock mass in the neighborhood of inclusions of quartz and augite.

In his Notes on Some Minerals of the Fichtelgebirge, Sandberger[20] gives analyses of *titanic iron* sand from the banks of the Eger, of *rhodonite* from Arzberg, of the *margarodite* covering orthoclase crystals in the druses of the lithionite granite of Epprechtstein, of the *chlorite* pseudomorphs of orthoclase crystals in the dolomite of Strehlenberg, and of a *lithium mica* from Fröstau, near Wunsiedel. The last named mineral is one of the constituents of a rock whose only other original component is white albite. Its analysis gave:

SiO$_2$	F	Al$_2$O$_3$	Fe$_2$O$_3$	MnO	MgO	K$_2$O	Na$_2$O	Li$_2$O	H$_2$O
50.11	1.36	1.36	1.01	1.01	.96	10.51	1.58	1.43	1.91

besides small amounts of TiO$_2$, SnO$_2$, FeO, CaO, CuO, As, Sb, Pb, Co, and B. The author thinks that there are certainly five distinct lithium micas known.

Katzer[21] mentions the occurrence of *arsenopyrite* and *quartz* crystals at Petrowitz, in Bohemia, of *sphalerite* and other sulphides, and of *siderite* at Heraletz, of *wollastonite* in fibrous masses on the contact of limestone with granite-gneiss, and of crystals of blue *cordierite* at Humpoletz, of andalusite at Cejod, of a calciferous *tourmaline* at Benitz, and of *gypsum* crystals at several localities in the same Kingdom. The tourmaline analyses gave:

SiO$_2$	Al$_2$O$_3$	B$_2$O$_3$	FeO	Fe$_2$O$_3$	MnO	CaO	MgO	Na$_2$O	K$_2$O	F	H$_2$O
35.53	30.73	5.59	5.67	7.67	1.17	3.16	2.82	4.38	.63	.12	2.86

Crystals of *epsomite* are described by Milch[22] from Stassfurt-Leopoldshall, Germany. They are implanted on a granular halite or a saliferous clay, and reach in many cases several centimeters in dimensions. They are all columnar in habit, and are remarkable for their richness in planes and for the marked character of their hemihedrism. The principal forms occurring in them are ∞ P$\overline{\infty}$, ∞ P$\underset{\infty}{\smile}$, ∞ P, ∞ P2, ∞ P2̄, P$\overline{\infty}$, and 2P$\overline{\infty}$.

Several rough, twinned crystals of *alabandite* from a deposit of the mineral in the Lucky Cuss Mine, Tombstone, Arizona, have been anal-

[20]Neues. Jahrb., f. Min., etc., 1892, ii, p. 37.

[21]Min. u. Petrog., Mitth, xii, p. 416.

[22]Zeits. f. Kryst., xx, p. 221.

yzed by Messrs. Moses and Luquer.[28] The mineral is of a dark, lead-gray color, with a brownish tarnish. *Wavellite* from the Dunellen Phosphate Mine, Marion Co., Fla., contains $Al_2O_3 = 37.076\%$, $P_2O_5 = 33.887\%$, and $H_2O = 26.366\%$.

Zincite crystals from Sterling, N. J., have again been analyzed. Grosser[24] finds in them $ZnO = 96.20$; $MnO = 6.33$; $Fe_2O_3 = .43$.

New Instruments.—A new signal for use in goniometrical measurements has been introduced to the notice of crystallographers by Goldschmidt,[25] which, it is believed, has several advantages over the Websky signal. A new adjusting apparatus for the goniometer has also been devised by the same crystallographer. It consists of an arm movable in four or five directions. By its use all the zones in a small crystal may be measured without the necessity of imbedding the crystal in wax more than once. A cheap heating apparatus to be used with the microscope has been constructed by Schrauf.[26] It is essentially a little box of a non-inflammable, poorly conducting material that is heated directly by a gas burner.

Staske[27] uses a very simple instrument for the production of curves of heat conductivity on mineral plates. It comprises a copper wire heated at one end and at the other touching the mineral slice, coated with paraffine.

Miscellaneous Notes.—Another investigation to determine the solubility of minerals in water under pressure, in the presence and absence of carbon-dioxide, has been made by Binder.[28] He finds that at 90° *bornite, chalcocite, marcasite, manganite* and *fluorite* are dissolved to an appreciable extent in pure water, and *cinnabar, cuprite,* and *pleonaste* to a slight degree only. When CO_2 is added to the solvent, *pyromorphite* dissolves, and *epidote* in small amounts. Under the same conditions *andalusite* and *anorthite* are decomposed.

The U. S. National Museum has issued a handbook of Geognosy, dealing with the materials forming the earth's crust. In it Mr. Merrill[29] outlines the characteristics of the aqueous, æolian, metamorphic

[28]School of Mines Quarterly, No. 3, xiii, p. 237.
[24]Zeits. f. Kryst., 1892, xx, p. 354.
[25]Zeits. f. Kryst., xx, 1892, p. 344.
[26]Ib., xx, 1892, p. 363.
[27]Ib., xx, p. 216.
[28]Min. u. Petrog., Mitth. xii, p. 332,
[29]Rep. of Nat. Mus. for 1890, p. 503.

and igneous rocks, and then describes briefly the principal members of each class.   The little book is well illustrated, and its contents are conveniently arranged for the student of the museum's collections.

All of the natural manganese oxides except pyrolusite and manganite yield red or violet solutions when digested with a mixture of sulphuric acid and water in equal proportions.[39]

[39]Thaddeef.   Zeits. f. Kryst., xx, 1892, p. 848.

From The American Naturalist, November 1st, 1892.

PARTIAL INDEX OF SUBJECTS.

INDEX OF AUTHORS.

SUMMARY OF PROGRESS

IN

Mineralogy and Petrography

IN

1893.

BY *H.W.T.*

W. S. BAYLEY.

FROM MONTHLY NOTES IN THE "AMERICAN NATURALIST"

PRICE 50 CENTS.

WATERVILLE, ME.,
GEOLOGICAL DEPARTMENT, COLBY UNIVERSITY,
1894.

No. 12

SUMMARY OF PROGRESS

IN

Mineralogy and Petrography

IN

1893.

BY

W. S. BAYLEY.

FROM MONTHLY NOTES IN THE "AMERICAN NATURALIST."

PRICE 50 CENTS.

WATERVILLE, ME.:
GEOLOGICAL DEPARTMENT, COLBY UNIVERSITY,
1894.

MINERALOGY AND PETROGRAPHY.[1]

The Origin and Classification of Igneous Rocks.—Mr. Iddings[2] has recently published at length the data upon which are based his conclusions concerning the causes of the different structures exhibited by the igneous rocks of Electric Peak and Sepulchre Mountain and of their varied mineral composition. The main results reached by this study have already been noticed in these pages.[3] It may be well again to call attention to the fact that in this region the different conditions attending the final consolidation of the ejected and of the intruded magmas affected not only their crystalline structure, but also their essential mineral composition; consequently, the molecules in a chemically homogenous fluid magma combine in various ways and form quite different associations of silicate minerals, producing mineralogically different rocks. For instance, biotite is an essential constituent of even the most basic of the intrusive rocks, while in the effusive phases it is rarely found in rocks containing less than 61% of SiO_2. Again, quartz is common in the coarser grained varieties of the former and is absent from those of the latter. Therefore, it is more proper to consider intrusive and effusive rocks that have a like chemical composition as *corresponding* or *equivalent* rocks, than those forms of the two series that have similar mineral compositions. The classification of igneous rocks should recognize the close dependence of structure and mineralogical composition upon geological relations. But, since the structure is the best exponent of these relations, structure should form the basis of this classification. Though giving most of his attention to the general subject of the relation existing between the structure and the geological position of the rocks of the area described, the author devotes a portion of his article to illustrating the intergrowths of hypersthene, pyroxene and hornblende that occur so plentifully developed in the rocks of the region.——In a second paper the same author[4] attacks the great problem of the origin of igneous rocks. He introduces the subject by outlining the growth of the theory first enunciated by Scrope, that the varieties of igneous rocks are the result

[1] Edited by Dr. W. S. Bayley, Colby University, Waterville, Maine.
[2] Twelfth Ann. Rep. Director U. S. Geol. Survey, Washington, 1892, p. 569.
[3] Cf. AMERICAN NATURALIST, April, 1890, p. 360.
[4] The Origin of Igneous Rocks. Bull. Philos. Soc. Wash., xii, 1892, p. 89.

of the differentiation of a homogeneous magma. Scrope's notion was a crude one, but it has been built upon little by little until it has, in the hands of Mr. Iddings, been placed upon a footing secure enough to warrant its being thoroughly tested by observation and experimentation. The author points out the evidences of the close relationships exhibited by the rocks emanating from a volcanic center and their differences from similar groups from other centers, and then takes up the question of the differentiation of molten magmas. He brings forward geological and chemical evidences of the fact of differentiation, and explains the act upon Soret's principle that in a solution whose parts are at different temperatures there will be a concentration of the salt in the colder parts. Lagorio has shown that rock magmas are solutions, and Iddings believes they are solutions of the chemical elements or of their oxides. Consequently, after differentiation has taken place and cooling sets in, different minerals are formed according to laws that depend upon the proportions of the oxides occurring in the differentiated portions. This is apparently contradictory to the view of Rosenbusch,[5] who regards rocks as having originated in the differentiation of a magma, but of a magma which is a solution of *silicate salts* in a *silicate solvent*. As a result of the condition of affairs suggested by Iddings the first eruption from a volcanic center would naturally possess a composition intermediate between those of succeeding eruptions. As a fact the author states that the sequence is usually a rock of intermediate composition, followed by less siliceous and more siliceous ones, to those very basic and very acid. The last eruptions are of very exceptional character. These will occur in small quantity only, and will be first eroded from the surface. Consequently these forms will be found principally in dykes. They are the forms to which Rosenbusch has given the group name "Ganggesteine." These rocks, according to Iddings, have their equivalents among volcanic flows, but the association of minerals in them is different. It is simply their structure, therefore, that characterizes the dyke rocks. They have originated in the same manner as have other eruptives, and consequently are not essentially different from them. The author's views are developed carefully and at considerable length. They will undoubtedly serve to turn the attention of petrographers to a subject that has lain neglected long enough—the comparative study of rocks of single geological provinces. The paper will well repay very careful reading by all petrographers and theoretical geologists, who should be

[5]AMERICAN NATURALIST, Nov., 1890, p. 1071.

glad to know that it is on sale by the Philosophical Society of Washington, from whose secretary it may be purchased for $1.

The Novaculites of Arkansas.—In his excellent discussion of the novaculites of Arkansas, Griswold[6] describes most of these rocks as consisting of very tiny irregular grains of quartz with occasional specks of carbonaceous matter. Originally the rock contained also well crystallized rhombohedra of calcite, traces of which are sometimes seen in the sections. Generally, however, the calcite has entirely disappeared, and its place is now occupied by a rhombic cavity, around which the quartz grains are packed as though they had been shoved about by the crystallizing carbonate. The good cutting qualities of Arkansas whetstones are thought to be due to the presence of these cavities. The purity of the Hot Springs novaculite is shown by an analysis that yielded:

SiO_2	Al_2O_3	Fe_2O_3	CaO	MgO	K_2O	Na_2O	Loss	Total
99.45	.26	.12	tr.	.19	.54	.06		=100.62

According to the author the rocks were first deposited as a mud or ooze, in which calcite crystallized. They were then consolidated by simple pressure, and finally, after upturning and erosion, they were supplied with a small quantity of secondary silica.

Petrographical News.—Osann[7] has discovered that the mineral heretofore regarded as sodalite in the Montreal eleolite-syenite is nosean, as it contains 5-6% of SO_4, and very little calcium. It is quite abundant in the rock, and is included as idiomorphic grains in its garnets. A microscopical test proposed by the author for distinguishing between nosean and sodalite is as follows: Moisten slide with dilute acetic acid to which a little barium-chloride has been added, and allow to stand in an atmosphere of the acid. Sodalite remains transparent, while nosean is covered with an opaque coating of barium sulphate.

The coloring matter of the black limestone of the Pyrenees is shown by Jannetaz[8] to be carbon, probably in the form of anthracite.

The new catalogue of geology and petrography issued by Ward's Natural Science Establishment, of Rochester, N. Y., deserves mention

[6] Ann. Rep. Geol. Survey of Ark. for 1890, Vol. iii, pp. 122–168.
[7] Neues Jahrb. f. Min., etc., 1892, i, p. 222.
[8] Bull. Soc. Franç d. Min., xv, 1892, p. 101.

in these notes because of the full list of rock names contained in it. The principal rock types are defined, and under each are given the technical names of all its varieties. It is further interesting as an indication of the growing importance of lithology in this country, since it is quite evident that Prof. Ward would not find it advisable to keep in stock such a large quantity of rock material were the demands for it rare. The catalogue may well serve the geologist as a table of petrographical synonyms.

A New Occurrence of Ptilolite.—A new occurrence of *ptilolite* has been discovered by Cross and Eakins[9] in Custer County, Col., about three miles southeast of Silver Creek, in the vesicles of a dull green devitrified pitchstone. The mineral is in very slender needles that are optically negative. An analysis made on very carefully selected material gave:

SiO_2	Al_2O_3	CaO	K_2O	Na_2O	H_2O	Total
67.83	11.44	3.30	.64	2.63	13.44	=99.28,

which is equivalent to R_7 Al_2 Si_{10} $O_{24}+6\frac{1}{2}$ H_2O, a formula identical with that determined for mordenite by Pirsson.[10] Clarke[11] regards a part of the water in each mineral as basic, and believes that mordenite, the ptilolite from Silver Creek and the original ptilolite (which is poor in Na_2O) are mixtures of the salts. Al_2 $(Si_2O_5)_5$ Ca H_2. 3Aq, Al_2 $(Si_2O_5)_5$ Ca H_2. 6Aq, Al_2 $(Si_2O_5)_5$ Na_2 H_2. 6Aq and Al_2 $(Si_2O_5)_5$ K_2 H_2. 6Aq.

Mineralogical News.—*Polybasite* and *tennantite* are reported by Penfield and Pearce[12] from the Mollie Gibson Mine in Aspen, Col. The former is the ore of the mine. It occurs massive, often associated with barite and siderite. It is of a grayish-black color, and has disseminated through it patches of the lighter tennantite. Analyses, corrected for impurities, follow:

	S	As	Sb	Aq	Pb	Cu	Zn	Fe
Polybasite.....	18.13	7.01	.30	56.90		14.85	2.81	
Tennantite...	25.04	17.18	.13	13.65	.86	35.72	6.90	.42

Crystals of both minerals are known to occur in several of the Colorado mines, though they have not yet been described.

[9]Amer. Jour. Sci., August, 1892, p. 96.
[10]Cf. AMERICAN NATURALIST, 1891, p. 372.
[11]Amer. Jour. Sci., August, 1892, p. 101.
[12]Amer. Jour. Sci., July, 1892, p. 15.

The *cerussite* from Pacaudière, near Roanne, Loire, France, is stated by Gonnard[13] to be associated with copper, silver and lead compounds, pyrite, limonite, quartz and calcite. Its simple crystals present a large variety of planes. Twinned crystals are common, and trillings are known. A description of the several types is given by the author. For sixty years past the same mineral has been known to occur at the argentiferous galena mines of Pontgibaud Puy-de-Dôme, but the fact has not been noted in the treatises on Systematic Mineralogy. All the crystals seem to have been formed at the expense of galena and bournonite by the action of CO_2 from the neighboring volcanic vents. The habit of its crystals is well described by Gonnard[14].

Morenosite [$(Ni\ Mg)\ SO_4 + 7H_2O$] in green stalactites from the foot of the Breithorn in Zermatt, yielded the same mineralogist[15] the figures $SO_3 = 28.7$; $NiO = 18.5$; $MgO = 6.5$; $H_2O = 46.5$. A single fragment of an ochre-yellow mineral from New Caledonia is a silicate of nickel, magnesium and iron:

SiO_2	Fe_2O_3	Al_2O_3	NiO	MgO	H_2O	Total
33.0	18.5	1.5	26.3	8.0	14.0	= 101.3

Frossard[16] substantiates the statement of Mallard that the black garnet *pyreneite* is a grossularite and not a melanite as reported by Raymond. Its density varies between 3.375 and 3.53.

Vesuvianite is reported by Pisani[17] from Settino in the Rhetian Alps. Its analysis gave:

SiO_2	Al_2O_3	FeO	CaO	MgO	MnO	Loss	Total
39.0	14.3	1.8	37.4	6.7	tr.	.9	= 100.1

The supposed *martite* crystals in the rock of Cuzeau, Mont Doré, are tabular hematites cemented into octahedra by magnesio-ferrite, as determined by Lacroix.[18]

In the basic clays of Condorcet near Nyons, Drôme, France, are boulders of siliceous limestone, with cavities whose walls are lined with bi-pyramidal *quartz* crystals, transparent *celestite*, *dolomite* and *calcite*. The quartz and celestite both contain rare planes beautifully developed.[19]

[13]Bull. Soc. Franç d. Min., xv, 1892, p. 35.
[14]Ib., xv, p. 41.
[15]Ib., xv, p. —.
[16]Ib., xv, p. 58.
[17]Ib., xv, p. 47.
[18]Ib., xv, p. 11.
[19]Ib., xv, p. 27.

Mineral Syntheses.—Bourgeois and Traube[20] having failed to produce *carbonates of the magnesium* group of elements by the reaction of urea, water and metallic chlorides on each other at 130° in sealed tubes, have made another attempt at their synthesis by substituting potassium cyanate for the urea. The attempt proved successful, needles of *aragonite* and rhombohedra of *dolomite* and *magnesite* having been produced under the conditions mentioned, when the chloride used was a mixture of the magnesium and calcium salts in molecular proportions.

By the slow action of dilute solutions of copper chloride upon freshly precipitated lead hydroxide at ordinary temperatures there is produced a blue powder consisting of octahedra and cubes of *percylite*, with which are associated quadruple twins of a colorless mineral supposed by C. Friedel[21] to be *phosgenite.*

Crocoite has been obtained by Ludeking[22] upon allowing a strong solution of caustic potash to stand for some time in contact with lead chromate in the presence of a little potassium chromate. By using a large excess of very strong caustic potash *phœnicochroite forms.* The crystallization of the latter substance is due to the abstraction of the solvent by the carbon-dioxide of the air, and of the former by a further reaction between the caustic potash and chromic acid.

New Minerals.—*Penfieldite.*—This mineral, discovered by Prof. Genth[23] on the slags from Laurion, Greece, is evidently produced by the action of sea water upon the materials of the slag. It is usually in the form of hexagonal prisms with basal planes, or in prisms tapered by pyramids. The color is white and the lustre vitreous to greasy. An analysis of the tapering crystals gave: $Cl = 18.55$, $Pb = 78.25$, $O = ———$, indicating the formula $Pb\ O.\ 2Pb\ Cl_2$.

Brazilite is a new tantalo-niobate from the iron mine Jacupiranga, in S. São Paulo, Brazil. Hussak[24] describes it as occurring in the magnetite-pyroxene rock called by Derby jacupirangite. It was separated by washing the decomposed residue of this rock in a miner's pan, and has heretofore been taken for orthite. Its crystallization is monoclinic with $a : b : c = .9859 : 1 : .5109.$ $\beta = 98°\ 45\tfrac{1}{2}'$. The forms observed in its crystals are $\infty P\overline{\infty}$, ∞P, $\infty P2'$, $—P\overline{\infty}$, oP, $P\overline{\infty}$,

[20]Ib., xv, 1892, p. 13.
[21]Ib., xv, 1892, p. 96.
[22]Amer. Jour. Sci., July, 1892, p. 57.
[23]Amer. Jour. Sci., 1892, p.
[24]Neues. Jahrb. f. Min., etc., 1892, ii, p. 141.

2P ∞', P and —P. The crystals are tabular parallel to the orthopin-
acoid and are nearly always twinned, frequently yielding very complicated groupings. The color of the larger crystals varies from sulphur-
yellow to black. Their hardness is 6.5 and density 5.006. The plane
of their optical axes is parallel to the clinopinacoid, and the double
refraction is negative. The extinction is 8°–15° in obtuse β, and the
pleochroism varies between dark-brown and oil-green. The minerals
associated with brazilite are apatite, magnetite, perofskite, ilmenite,
and a spinel. An analysis of the new minerals is promised shortly.

Landauer's Blowpipe Analysis.—This little book[*] will be
cordially welcomed by English and American teachers in colleges in
which the use of a large manual of blowpipe analysis is undesirable.
It is as suitable for classes in mineralogy as in chemistry, since it will
enable the student to determine the composition of a mineral as rapidly
as will the use of the great majority of Determinative Mineralogies
upon the market. Moreover, it possesses one desirable advantage over
those schemes in which the hardness, color and streak of chemical
compounds are made to serve as distinctive tests for them, in that it
compels the experimenter to study the chemical nature of the substance with which he is working. A mineral is a definite chemical
substance. A student of mineralogy who is unfamiliar with the composition of bodies with which he is working, though he may know
considerable about their physical properties, is neglecting the founda-
tion upon which his knowledge of minerals must rest. The little book
before us is an excellent introduction to the larger works like those
of Brush and Plattner. It is, besides, complete enough for most of
the purposes to which such a book is usually put. Beginning with a
good description of the apparatus and reagents necessary to blowpipe
manipulation, it follows with an account of the operations employed,
describes Bunsens flame reactions, mentions the distinctive tests for the
various chemical elements, gives Landauer's and Egleston's schemes
for the systematic examination of inorganic substances, and closes
with tables exhibiting the reactions of the various metallic oxides,
and in a condensed form the results of the different operations
described in the text. The book must find a place in many labor-
atories.

[*]Blowpipe Analysis, by J. Landauer. Authorized English Edition by
James Taylor. Second Edition. Macmillan & Co., 1892, pp. 14 and 173.

From The American Naturalist, January 1st, 1893.

MINERALOGY AND PETROGRAPHY.[1]

The Rocks of the Thalhorn.—In the Thalhorn of the Upper
Amariner Thal are found a porphyritic granite, between conglomerates
composed of gabbro pebbles in a schistose matrix, and also serpentines, massive gabbro, schists, and various contact rocks. Linck[2] gives
a good petrographical description of all these, and geological notes of
their occurrence. The main granite mass is a portion of the well-
known Kamm granite. It is found in dykes and flows, and it varies
in its composition and structure from a typical granitite containing
two feldspars, through porphyritic granite and syenite to lamprophyric
minettes. The unaltered sediments near the eruptive are graywackes.
On the contact with the granite the clastics are altered to knotty schists
that are predominantly biotite schists flecked with light spots, consisting mainly of quartz and feldspar in micropegmatitic intergrowths,
surrounded by biotite. Extreme alteration gives rise to hornstones, of
which the writer recognizes several varieties. In these biotite, feldspar, hornblende and micropegmatite are so orientated as to resemble
the poicilitic structure of many diabases and other basic rocks. Hornblende is abundant in them as needles scattered through the groundmass and as large phenocrysts. The conglomerates occupy the greater
share of the writer's attention. In one group acid pebbles occur in a
sandy or clayey matrix of basic detritus, in which biotite, feldspar and
hornblende are new products of alteration. A second group includes
rocks made up partly of gabbro material. Here the author again
recognizes two groups, in one of which diallage and other gabbro constituents are occasionally present in the groundmass, and a second in
which gabbro material forms a very large portion, either of the matrix
or of the pebbly portion of the rock. In either case the rock is much
altered, with the resulting formation of plagioclase and hornblende.
The serpentine of the region was originally an olivine-enstatite rock
and not a gabbro as has been supposed.

The New Jersey Eleolite-Syenite.—The New Jersey Eleolite-
syenite dyke described by Emerson[3] is again studied by Kemp,[4] who

[1] Edited by Dr. W. S. Bayley, Colby University, Waterville, Me.
[2] Mitth. d. geol. Landesanst v. Elsass-Loth., iv, 1892.
[3] Amer. Jour. Science, iii, xxiii, p. 302.
[4] Trans. N. Y. Acad. Sci., Vol. xi, p. 60.

declares that the earlier description applies only to that phase of the rock occurring in the northern and the southern portions of its extent. The pyroxene throughout the dyke is aegerine. Cancrinite and soda-lite are both fairly abundant in it. An analysis of specimens collected from about the point visited by Emerson gave:

SiO_2	Al_2O_3	Fe_2O_3	MnO	CaO	MgO	K_2O	Na_2O	Loss	Total
50.36	19.34	6.94	.41	3.43	not det.	7.17	7.64	3.51	=99.80

Eleolite porphyries with a tinguaitic groundmass are closely associated with the more abundant syenite, and along the eastern side of the great dyke are smaller ones of ouachitite and fourchite. The basic material of these small dykes, when first[5] studied, was regarded as porphyrite. Contact effects produced by the intrusion of the syenite through the surrounding shales are noticed on the east side of the dyke, where the sedimentaries have been changed to biotitic hornfels.

Mica Peridotite from Kentucky.—A mica peridotite[6] from a dyke in Crittenden Co., Ky., is composed essentially of biotite, serpentine, and perofskite, with smaller proportions of apatite, muscovite, magnetite, chlorite, calcite, and other secondary products. The biotite and serpentine constitute about 75% of the entire rock. The mica is in large plates in which are scattered the grains and shreds of serpentine. The composition of the rock follows:

SiO_2	TiO_2	Al_2O_3	Fe_2O_3	FeO	CaO	MgO	K_2O	Na_2O	H_2O	P_2O_5	CO_2
33.84	3.78	5.88	7.04	5.16	9.46	22.96	2.04	.33	7.50	.89	.43

and small quantities of Cr_2O_3, MnO, NiO, CoO, BaO and Cl. The rock represents a new type of peridotite in which biotite takes the part of an amphiboloid in the more usual types.

Rhyolites in Maryland and Penn.—G. H. Williams[7] has identified an extensive series of old volcanic rocks in the South Mountain region of Pennsylvania and Maryland. The rocks have hitherto been considered sedimentaries, but to the writer they exhibit all the peculiarities of eruptives, though some of the beds are fragmental tufas and breccias. The two principal types are rhyolite and basalt. The former possesses all the features of recent eruptives, such as flowage

[5] Amer. Jour. Sci., III, xxxviii, p. 130.

[6] J. S. Diller, Amer. Jour. Sci., xliv, 1892, p. 286.

[7] Ib., xliv, 1892, p. 482.

19

lines, spherulites, lithophysae and amygdaloidal cavities. Quartz and
an alkaline feldspar are the prevailing phenocrysts, while the groundmass is a quartz-feldspar mosaic. The basalts are much altered, but
their structure is clearly that of an eruptive. A detailed account of
the rocks is promised later.

The Nepheline and Leucite Rocks of Brazil.—A more
careful study of a few of the Brazilian nepheline and leucite rocks
undertaken by Hussak[6] has resulted in the discovery of leucite in some
of the phonolites, and in the detection of leucite-tephrites containing
pseudo-crystals. The leucitophyres consist of phenocrysts of sanidine,
augite, nepheline and pseudo-leucites in a groundmass of small zeolitized leucites, augite, magnetite and nepheline. The leucite-tephrites
are all characterized by the possession of the pseudo-leucites. In many
cases these are nothing but spherical masses of the rock material surrounded by biotite plates. In other cases the biotite surrounds analcite or mixtures of analcite and calcite. The structure of several of
these rocks is the diabasic. With these the author would place a rock
described by Eigel[9] from the Cape Verde Islands, and the augite-por-
phyrite described by Kemp[10] from Deckertown, N. J., in both of which
traces of leucite are thought to have been discovered. Hussak has
also found a leucitite dyke in phonolite near Poços de Caldas, and a
leucitite tufa composed of fragments of basalt, isolated crystals of
leucite changed to analcite, pieces of augite and crystals of magnetite.
The author concludes his paper with remarks on 'pseudo crystals'
combating the view of Derby that they are true leucite crystals filled
with inclusions of the rock's groundmass.

The last named writer[11] has examined the Peak of Tingua with some
care, finding eleolite-syenite, phonolite and dykes of basic rocks. The
syenite and phonolite are thought to be phases of the same magma, as
they apparently grade into one another. The phonolitic phase occurs
both in dykes and in flows associated with phonolite tufas. The origin
of the pseudo-crystals is discussed briefly.

Petrographical News.—Brauns[12] has discovered hauyne in the
pumice sandstone near Marburg, a mineral hitherto unobserved in the

[6]Neues. Jahrb. f. Min., etc., 1892, II, p. 141.
[9]AMERICAN NATURALIST, Feb., 1892, p. 165.
[10]See above under 'The New Jersey Eleolite-Syenite.'
[11]Quart. Jour. Geol. Soc., May, 1891, p. 251.
[12]Zeits. d. deutsch. geol. Gesell., xliv, 1892, p. 149.

rock because of the loss of its characteristic blue color through alteration. The list of minerals common to this rock and to those of the Laacher See is now complete, so that the belief in a common origin for them is rendered almost a certainty.

C. W. Hall[13] gives a few notes on rocks collected from Central Wisconsin, describing very briefly hypersthene and quartz gabbros in which there is much secondary hornblende, and quartz diorites and gneisses regarded as squeezed gabbros.

A fourchite boulder in which are large arfvedsonite phenocrysts is mentioned by Kemp[14] as occurring at Aurora, Cayuga Co., N. Y. The same author mentions the existence of rhyolite, hypersthene, andesite and andalusite-hornstone from near Gold Hill, Toole Co., Utah.

Spherulites[15] of andalusite occur in the carboniferous clastic schists of Beaujeu, France. The schists are composed of black and white mica fragments in a paste of sericite and hematite.

Turner[16] makes brief mention of basaltic, andesitic and rhyolitic lavas, whose source was the late Tertiary cone Mt. Ingalls, in California.

Crystallographic Study of Diopsides.—Some very careful crystallographic observations have been made by A. Schmidt[17] upon the diopsides of the Alathal, of Achmatowsk, of Nordmark, of the Zillerthal and the Arany-Berg. Many crystals from each of these famous localities were examined, and much new data was obtained concerning the mineral. The following new planes were discovered: $4P\bar{2}$ and $5P\frac{4}{3}$ on the white diopside from Achmatowsk; $\frac{1}{2}P\bar{\infty}$ on the green variety from the same place; $\infty P6'$ in the Nordmark species; $\infty P\overline{10}$, $\infty P4'$ and $\infty P\frac{4}{7}$ on the nearly colorless small crystals from Schwarzenstein in the Zillerthal, and $\infty P\bar{7}$ and $P\bar{4}$ on the black Arany-Berg mineral. The form $P\bar{4}$ appears in Goldschmidt's 'Index,' but no reference to it could be found by the author in the original memoirs. The axial ratios of the different varieties are:

Alathal....................................	$1.0895 : 1 : .5894 \quad \beta = 74°15'47''$
Achmatowsk (white)................	$1.0909 : 1 : .5899 \quad \beta = 74°10'42''$
Achmatowsk (green)..............	$1.0951 : 1 : .5985 \quad \beta = 73°31'8''$

[13]Minn. Ac. Nat. Science, III, No. 2, p. 251.

[14]Trans. N. Y. Acad. Sci., xi, p. 92.

[15]Lévy. Bull Soc. Franç d. Min., xv, 1892, p. 121.

[16]Amer. Jour. Sci., Dec., 1892, p. 455.

[17]Zeits. f. Kryst., xxi, 1892, p. 1.

Nordmark................................ $1.0915 : 1 : .5848$ $\beta = 74°38'59''$
Zillerthal (colorless)................... $1.0922 : 1 : .5887$ $\beta = 74°16'28''$
Arany-Berg (yellow)................... $1.0945 : 1 : .5918$ $\beta = 74°19'38''$
Arany-Berg (black)........... ...... $1.0913 : 1 : .5875$ $\beta = 74° 4'53''$

The optical angle for the Nordmark crystals is $2Vna = 60°44'$, and $C \wedge c = 45°21'$. For the dark Zillerthal diopside $2Vna = 58°56'$ and $C \wedge c = 34°4'$.

Herderite from Hebron, Maine.—A single specimen of Herderite from Hebron, Maine, is described by Wells and Penfield[18] as a few yellowish white crystals on albite. The crystals have a tabular habit, with oP, ∞ P, 3P and ⅓P the only forms observed. The density is 2.975 and composition :

P_2O_5	BeO	CaO (by diff.)	H_2O	F	Insol.	Total
40.81	15.32	32.54	5.83	.40	5.27	= 100.17

Corresponding to Ca Be (OH) PO_4, or a herderite in which nearly all of the fluorine is replaced by hydroxyl.

Mineralogical Notes—A *calcium carbonate* of secondary origin from the Marble Mountains of Wolmsdorf in Glatz has been analyzed by Kosmann[19] with the following astounding result: Ca $CO_3 = 4.32$; chemically combined $H_2O = 1.54$; mechanically combined $H_2O = 94.13$. The author believes the mineral to be a hydrated carbonate $CaCO_3 + 2H_2O$ capable of absorbing a large quantity of water, similar to the 'Mountain Milk' of Rose.

The *friedelite* of the Manganese mine of Sjogrube, Orebro, Sweden, occurs in large quantity in clefts, veins, etc., that are partially filled with calcite. An analysis yielded Igelström :[20]

SiO_2	Cl	MnO	FeO	CaO	MgO	NaO	H_2O	Total
34.36	3.00	45.88	1.35	1.50	1.50	2.79	9.00	= 99.38

On the *Azurite* from the Laurion Mts., Greece, Zimanyi[21] has found 28 forms, three of which ($\tfrac{3}{4}P\bar\infty$, $\tfrac{3}{4}P\check\infty$ and $\tfrac{3}{4}P\overline{\infty}$) are new. The crystals have the usual habit of the mineral, and they compare favorably in beauty with those from Chessy, Arizona and Utah.

[18]Amer. Jour. Sci., xliv, 1892, p. 114.
[19]Zeits. d. deutsch. geol. Ges., xliv, 1892, p. 155.
[20]Zeits. f. Kryst., xxi, p. 92.
[21]Ib., xxi, p. 86.

Tremolite[21] pseudomorphs after sahlite from the limestone of Canaan,
Ct., have the composition:

SiO₂	Al₂O₃	Fe₂O₃	FeO	CaO	MgO	K₂O	Na₂O
60.98	.10	.12	.19	14.64	23.62	.13	.21

The controversy over the nature of *Melanophlogite* is not yet ended.
Bombicci[22] has recently defended himself against the attack of Friedel,
and in his defense he accuses his opponent with misquoting him.

New Minerals.—*Ganophyllite*, from Harstige, near Pajsberg,
Sweden, is a manganese zeolite[24] that is associated with barite, lead,
and rhodonite. It occurs in large brown monoclinic, prismatic crystals, in which ∞ P is combined with the base and the clinodome. $a : b$
$: c = .413 : 1 : 1.831.$ $\beta = 86°39'$. On cleavage plates parallel to oP
a percussion figure may be produced, one of whose rays is parallel to a
and the other two inclined at 60° to this. Plane of optical axes is
perpendicular to ∞ P∞, with c the first bisectrix. 2E (air) = 41°53′
for sodium light, and 2V = 23°52′. The pleochroism is strong $c = A$
= yellow-brown; $a = B$ and $b = C$ = colorless. The density is
2.84 and hardness ≐ 4. A mean of two analyses gave Hamberg a
result that may be represented by $8SiO_2$, Al_2O_3 $7MnO + 6H_2O$.

SiO₂	Al₂O₃	Fe₂O₃	MnO	CaO	MgO	PbO(?)	K₂O	Na₂O	H₂O	Total
39.67	7.95	.90	35.15	1.11	.20	.20	2.70	2.18	9.79	= 99.85

Pyrophanite, described by the same author as occurring in the same
mine, is a manganese titanium compound isomorphous with ilmenite.
An analysis gave:

SiO₂	TiO₂	MnO	Fe₂O₃	Sb₂O₃	Total
1.58	50.49	46.92	1.16	.48	= 100.63

It is found as brilliant, deep red, transparent tables, associated with
ganophyllite. $a : c = 1 : 1.369$. The double refraction is strong, and
the indices of refraction for sodium light are $\omega = 2.481$, $\epsilon = 2.21$.
Density is 4.537.

Synthesis of the Members of the Sodalite Group.—The
minerals of the sodalite group have been manufactured by Morozie-

[21] W. H. Hobbs. Amer. Geol., July, 1892, p. 44.
[22] Bull. d. l. Soc. Franç d. Min., xv, p. 144.
[24] Ref. Neues. Jahrb. f. Min., etc., 1892, II, p. 284.

wics[25] as microscopic crystals. A mixture of 65 parts $SiO_2 + 3Aq$, 44 parts $Al_2O_3 + 3Aq$, and 33 parts gypsum, heated in a platinum crucible with an excess of Glauber's salt, yielded tiny cubes and dodecahedra of hauyne or sodalite. When heated with an excess of $Na_2SO_4 + Na Cl$ a substance was obtained that is supposed to be an isomorphous mixture of the two minerals above mentioned, and in addition some sodalite crystals were produced. When heated with $Na Cl$ alone sodalite only resulted.

Methods and Instruments.—A simple method for determining the value of the optical angle in thin sections of minerals is described by Lane.[26] It consists essentially of the measurement of the angular distance between the hyperbolas of the biaxial interference figure by means of the sub-stage mirror.

A cheap form of crystal refractometer constructed on the same principles as the larger Zeiss instrument has been made by Czapske.[27] The height of the complete instrument is only 25 cm. It is suitable for all ordinary refraction work.

An Appendix to the " Gems of North America."—Mr. Kunz has issued an appendix to his valuable 'Gems and Precious Stones of North America'[28] that brings the volume up to date. Most of the material in the appended chapter has appeared in the journals, but some of the information it contains is new. The author states that the sapphire gravels of Ruby Bar, Montana, and the turquoise mines of New Mexico are now being worked by companies that expect their outlay of capital justified by a goodly yield of gem material. The turquoise company has already taken from their diggings about a hundred thousand dollars worth of gems.

[25]Neues. Jahrb. f. Min., etc., 1892, II, p. 139.

[26]Science, Dec. 23, 1892, p. 354.

[27]Neues. Jahrb. f. Min., etc., 1892, I, p. 209.

[28]Cf., AMER. NATURALIST, Dec., 1891, p. 1119.

From The American Naturalist, March 1st, 1893.

MINERALOGY AND PETROGRAPHY.[1]

Description of the New Rock Type, Malchite.—The new rock, malchite, referred[2] to a few months ago as the granitic dyke form of diorite, is now described in some detail by its discoverer, Osann.[3] It forms dykes cutting granite in the Odenwald, Germany. In a dense groundmass are rare phenocrysts of dark mica, pale green plagioclase and quartz. The mica is biotite and the plagioclase labradorite. In addition to these the microscope reveals the presence of idiomorphic green hornblende, allanite and sphene. The groundmass in which these lie strongly resembles that of some tinguaites, with hornblende and quartz in place of aegerine and nepheline. It consists of a fine granular aggregate of feldspar and quartz, the latter with occasional idiomorphic contours, and prisms of hornblende imbedded in the aggregate, the prisms often arranged in flowage lines. An analysis of a fresh specimen of the rock yielded:

SiO_2 Al_2O_3 Fe_2O_3 FeO MgO CaO Na_2O K_2O H_2O SO_4 P_2O_5 Total
63.18 17.03 .24 6.37 .92 4.17 4.44 2.91 .52 .19 .23=100.20

The Petrography of Hokkaido, Japan.—In a general geological sketch of Hokkaido, (Jezo or Yesso), Japan, Jimbo[4] declares that the island consists largely of paleozoic beds, probably underlain by amphibolites and various other schists, and cut by granite, diorite, gabbro, peridotite, and serpentine. In the lower portion of the paleozoic the beds consist largely of pyroxenites, with traces of radiolarian remains, phyllites, quartz-schists, limestone, and serpentine. The pyroxenites are aggregates of light colored augite, quartz and feldspar, in which the augite is often more or less changed to epidote and glaucophane. Where the granite cuts the clastics the clay slate is changed by contact action to a biotitic clay slate, to hornfels and to mica schist, with the latter nearest the eruptive. Tourmaline occurs in the schist and cordierite in this rock and in the mica slate. An amphibolite in the contact belt is supposed to be an altered tufa. Schistose granites, diorites and gabbro are phases of the corresponding

[1]Edited by Dr. W. S. Bayley, Colby University, Waterville, Me.
[2]AMERICAN NATURALIST, May, 1892, p. 422.
[3]Mitth. Gross. Bad. geol., Landesanst ii, p. 380.
[4]General Geological sketch of Hokkaido, with special reference to the petrography. Satporo, Hokkaido, Japan, 1892.

massive rocks associated with the contact products. Diabases occur as sheets in the unaltered paleozoic beds, and serpentines derived from gabbros and from dunites are met with cutting these at various localities. In addition to paleozoic there are also tertiary rocks on the island, and these are cut by their own systems of dykes and bosses, and are interbedded with characteristic sheets of lava, and layers of tufas. The tertiary volcanic rocks are pyroxene and hornblende andesites, propylites and rhyolites. The pyroxene andesites contain both orthorhombic and monoclinic pyroxenes and occasionally some olivine. They have also a glassy base which sometimes becomes so abundant as to 'resemble pumice. The hornblende andesite is strongly porphyritic with large phenocrysts of hornblende. The rhyolites are both compact and glassy, in which latter case they are vesicular.

Two Peculiar Rocks from Siberia.—Two very remarkable rocks are described by von Chrustschoff[5] from Taimyr-Land, Siberia. One is an ophitic aggregate of anorthoclase and nosean, containing as accessories sanidine, plagioclase, amphibole, biotite, melanite, magnetite, sphene, zircon and glass. The anorthoclase is in long, narrow crystals of the following composition:

SiO_2	Al_2O_3	Fe_2O_3	CaO	MgO	K_2O	Na_2O	Total
64.59	19.84	2.24	1.26	.63	3.53	7.88	=99.97

Corresponding to Or_2, Ab_6, An_1. The feldspar is usually idiomorphic with respect to the nosean, whose period of formation was between that of the biotite and that of the hornblende. The nosean is in very large quantity. Its density is 2.266 and composition:

SiO_2	Al_2O_3	Fe_2O_3	Na_2O	K_2O	CaO	H_2O	Cl	SO_3	Total
37.83	26.59	.38	22.40	1.63	.54	.87	1.66	8.68	= 99.98

The zircon is of the trachytic type, and is the only accessory of any importance. The author calls the rock taimyrite. The second rock is composed of anorthoclase, sanidine, biotite, and amphibole as essential components, and the other minerals mentioned above in connection with taimyrite as accessories, except that sodalite here replaces nosean. The zircon is of the granitic type, and the rock possesses the granitic texture.

[5] Bull. d. l'Acad. Imp. des Sciences St. Petersb. Mél. Geol. et Paleont., i, p. 153.

An Ottrelite Bearing Conglomerate in Vermont.—It is not uncommon to find ottrelite forming 25% of the schistose groundmass of the conglomerate[6] at the base of the Lower Cambrian, near Rutland, Vermont. The same mineral occurs along shear planes in a blue quartzite and constitutes 40% of a massive bed of the conglomerate. In the last named rock the ottrelite is in rudely circular areas, lying in a dark colored quartz. The areas consist usually of radiating plates of the mineral, disposed in a single plane. Its commonest inclusions are quartz and feldspar, while sericite often forms the centers of the radiating bundles. In the latter case the ottrelite is oriented in parallel position with the mica. Other inclusions within the ottrelite besides those above mentioned are crystals of zircon and rutile, flakes of graphite and plates of ilmenite. In other cases the ottrelite is in plates including large areas of the groundmass of the rock, which is a granulated mixture of quartz and albite (?) in about equal proportions, a large quantity of sericite, and some biotite. In this groundmass associated with rutile are crystals and plates of anatase. No traces of its original clastic structure remain in the rock, though its conglomeratic character is beyond dispute.

Lithophysæ in the Rocche-Rosse.—In parts of the Rocche-Rosse lava stream of Monte Pelato, Lipari, are spherulites with lithophysal characteristics. In some specimens examined by Cole and Butler[7] the spherulitic growth originated about the walls of steam vesicles, and progressed outward into the rock; in other cases they grew inward until they have completely filled the space that was formerly vacant. The importance of the paper lies in the fact that it acknowledges the correctness of many of Idding's views with respect to the formation of lithophysæ, and contradicts the view that regards all hollow lithophysæ as the result of the decomposition of spherulites.

The Composition of the Dune Sands of the Netherlands.—A very elaborate paper by Retgers[8] on the constitution of sand composing the dunes on the west coast of Holland at Sheveningen, near the Hague, contains a large amount of information concerning the character of sands and the method of determining the nature of their constituents. The author carefully fractioned large quantities of the dune sand by the ordinary methods of fractional precipitation in

[6]C. L. Whittle, Amer. Jour. Sci., Oct., 1892, p. 270.
[7]Quart. Jour. Geol. Soc., xlviii, 1892, p. 438.
[8]Recueil des Travaux Chimiques des Pays-Bas., xi, 1892, p. 169.

the usual heavy liquids and by means of the molten substances suggested by himself[9] for this purpose a few years ago, thus obtaining
mixtures of mineral grains of about the same density. These then
were studied carefully by comparison of their indices of refraction, by
immersing them in liquids of known optical densities, until one was
found in which the grains became almost invisible. The index of
refraction of these is nearly that of the liquid, consequently their
nature is thus approximately determined. Microchemical tests and
the ease with which cleavage laminæ were produced, served to distinguish accurately between minerals having nearly the same refractive
index. The principal minerals identified by the author are orthoclase, quartz, microline, plagioclase, cordierite, calcite, apatite, amphibole, tourmaline, pyroxene, epidote, sphene, sillimanite, olivine, garnet, staurolite, disthene, corundum, spinel, rutile, zircon, magnetite and
ilmenite. The surprising discoveries are those of cordierite, calcite
and olivine, and of glaucophane among the amphiboles. The proportions of the various minerals present according to specific gravity was
2.5% between 2.5 and 2.6; 85% between 2.6 and 2.7; 7.5% between
2.7 and 3. ; 1.5% between 3 and 3.3; 1% between 3.3 and 3.6; 2.4%
between 3.6 and 4.2; .1% between 4.2 and 5.2. The sands are supposed to have come mainly from the rocks of archean terraces.

Quartz-Gabbro in Maryland.—In the Baltimore gabbro area,
according to Grant,[10] are quartz gabbros consisting of bytownnite,
quartz, hypersthene, secondary hornblende, and a few accessories.
The quartz is limpid, and is almost free from inclusions, except for
lines of small liquid cavities that traverse the grains, as is usual in
granitic quartz. Diallage, which is so common in the normal gabbro
of the region, is entirely absent from the quartz-bearing phases, which
thus becomes a quartz norite.

Minerals from the Diamond Fields of Brazil.—Hussak[11]
describes the characteristics of crystals of brookite, cassiterite and
xenotime from the diamond region of Dattas, Minas Geraes, Brazil.
On *brookite* from the sands of Diamantina was found the new pyramid,
$\frac{1}{4}P\tilde{2}$. The *cassiterite* is from Manquinho, near São Paulo. It occurs
in a rubellite-bearing lepidolite granite. The *xenotime* accompanies
the brookite in the sands of Dattas. On one doubly terminated crys-

[9]Cf., AMER. NATURALIST, 1890, p. 175.
[10]Johns Hopkins Univ. Circ. No. 108.
[11]Min. u. Petrog., Mitth. xii, p. 455.

tal were found the two new pyramids $\frac{4}{3}$P and $\frac{1}{3}$P. The axial ratio of these crystals is 1 : .61775. The author has also made a crystallographic examination of the *monasite* occurring so abundantly in the Brazilian granites and gneisses. The crystals of this substance are always tabular parallel to ∞ P$\overline{\infty}$. They contain the same forms as do the Ilmengebirge crystals, but are never twinned. , Upon washing a portion of sand from Bohia a 3 mm. long crystal of *euclase* was obtained whose density is 3.1. It is very rich in planes, being possessed of not less than three prisms, six clinodomes and three negative pyramids, beside the clinopinacoid.

Mineralogical Notes.—*Christianite* crystals are reported by Gonnard[12] as lining geode cavities in the basalts of dykes at Queyrières and Fay le Froid, Haute Loire, France. In the latter case the christianite groups enclose many crystals of augite. The trachyte of Montcharet, occurring as a dyke in granite, is cut by fissures whose walls are lined by *chabasite*.[13] The cubic faces of *galena* crystals implanted in druses of quartz at Pontgibaud are roughened by little cavities whose walls have the positions of octahedral planes. The phenomenon is regarded by Gonnard[14] as the result of corrosion. The same author mentions the existence of large crystals of *beryl* in the granites of Droiturier, near La Palisse, Allier, *psilomelane* in mammillary forms at Croix Moraud, Mt. Doré, and cubic pseudomorphs of *quartz* after some unknown mineral, probably fluorite, in the vicinity of d'Aubenas, Ardèche.

Three specimens of *melilite* from Mt. Somma, with densities of 2.917, 2.932 and 2.945 respectively, were powdered, purified, and analyzed by Bodländer[15] with this result:

SiO_2	Al_2O_3	Fe_2O_3	CaO	MgO	K_2O	Na_2O	H_2O	Total
41.34	10.37	4.29	33.84	5.79	1.13	3.45	.08	= 100.29

The author combats the view of Vogt that melilite is an isomorphous mixture of the gehlenite and akermanite molecules. He thinks that the negative variety is an admixture of $R''SiO_2$ and the aluminate R'''_2O_4R'', while the positive variety is a compound of the same silicate with the aluminate $R_2'''O_4R_2''$. Intermediate varieties are isomorphous mixtures of these.

[12]Bull. Soc. Franç d. Min., 1892, xv, p. 28.
[13]Ib., p. 31.
[14]Ib., p. 34.
[15]Neues. Jahrb. f. Min., etc., 1893, i, p. 15.

Moses[16] records the analysis of a granular *nickel arsenide* associated with native silver and siderite in a mine 18 miles west of Silver City, N. Mex. The silver is imbedded in arborescent forms in the brittle gray nickel ore, and this in turn is in a gangue of siderite. The analysis made on impure substance gave:

SiO$_2$	Pb	Ag	As	Ni	Co	Fe	Total
4.56	tr.	8.33	67.37	11.12	5.13	2.64	= 99.20

Regarding the SiO$_2$ and Ag as impurities the composition takes a form that may be represented by RAs$_2$ in which R = $\frac{1}{2}$ Ni $\frac{1}{3}$ Co and $\frac{1}{6}$ Fe, corresponding to a *nickel skutterudite*.

On crystals of topaz from the Province of Omi and from the tin mines of Yenagari Mino, Japan, Matthew[17] finds four pyramids, seven prisms, one of which, ∞ P$\bar{\gamma}$, is new, the three pinacoids, three brachydomes and two macrodomes.

Optical Anomalies.—After an exceedingly careful examination of many sections of appophyllite crystals and a comparison of the phenomena they present with those presented by combinations of thin biaxial plates placed one upon the other, Klein[18] concludes that the mineral in its geometrically tetragonal crystals is an intimate mixture of optically positive and optically negative triclinic lamellæ. The positive constituent seems to differ from the negative element in containing no crystal water, since upon heating the positive component appears to increase in quantity. Negative appophyllite becomes positive upon loss of 4½ molecules of crystal water. The investigation is a beautiful piece of accurate optical work.

In a reply to Mallard's[19] remarks on the black garnet pyrenaite Brauns[20] states that the structure described by the first mentioned author is exactly what should be expected of a dodecahedral substance under strain, and that the peculiarities of this garnet's optical properties may be easily explained on the Klein-Brauns theory of strain.

Upon soaking in oil sections of zeolites that have been rendered cloudy by loss of water, they again become sufficiently transparent for the study of their optical properties. Rinne[21] has taken advantage of

[16]School of Mines Quart., xiv, No. i, p. 49.

[17]Ib., xiv, No. 1, p. 53.

[18]Neues. Jahrb. f. Min., etc., 1892, II, p. 165.

[19]AMERICAN NATURALIST, Oct., 1892, p. 849.

[20]Neues. Jahrb. f. Min., etc., 1892, I, p. 217.

[21]Ref. Neues. Jahrb. f. Min., etc., 1892, II, p. 237.

this phenomenon and has carefully examined a number of the members of the group with a view to learning something of the changes effected in them by the loss of water. *Natrolite* appears monoclinic under these conditions, *scolecite* orthorhombic, *stilbite* orthorhombic, and each of the other zeolites affords a corresponding meta-zeolite. The optical anomalies often observed in these minerals is thought to be undoubtedly due to partial loss of water.

Isomorphism.—After a long mathematical discussion of the theory of the structure of isomorphous mixtures and upon comparison of the results of investigations upon the optical properties of mixed crystals, Poeckel[22] concludes that we have not yet sufficient data to decide as to whether Mallard's lamellæ theory of the constitution of these bodies is correct or not.

By the use of the method[23] in which colored and colorless crystals of supposed isomorphous substances are allowed to form under the microscope Retgers[24] has proven that the alkaline ferrates are isomorphous with the corresponding sulphates, selenates, molybdates and tungstates, that the potassium tellurates and osmiates are isodimorphous, and that the rutheniate of this metal is isomorphous with its uranate.

The arguments for and against the view as to the isomorphism of calcite and dolomite are given respectively by Brauns and Retgers[25] in a recent letter to the Neues, Jahrbuch. The discussion is too involved to warrant an intelligible abstract in these notes.

Etched Figures.—The matrix of the African diamonds is capable of resorbing[26] *diamonds*, producing on their faces irregular, long, and hemispherical hollows, associated with which are little spheres and grains of black carbonaceous substance, supposed to be a compound of iron and carbon.

Hofer[27] describes corrosian forms on the *calcites* of Steierdorf, Banat, and of Rauris and Salzburg, and ascribes the hexoctahedral faces ⅙O2 on the *fluorite* of Sarnthal, Tyrol, to corrosive processes.

[22]Neues. Jahrb. f. Min., etc., B. B., viii, 1892, p. 117.
[23]AMERICAN NATURALIST, June, 1892, p. 517.
[24]Zeits. f. Physik. Chem., x, 5, 1892, p. 529.
[25]Neues. Jahrb. f. Min., etc., 1892, II, p. 210.
[26]Ber. deutsch. chem. Ges., 1892, p. 2470.
[27]Min. u. Petrog., Mitth. xii, p. 487.

From The American Naturalist, April 1st, 1893.

Microchemical Reactions.—The methods of testing for traces of ammonia under the microscope, and of precipitating metals with H₂S are described in a few words by Streng.[28]

Directions for the detection of the following minerals in small particles are given by Lemberg:[29] *Scapolite, hauyne, sodalite, eudialite, lazurite, sulphur, olivenite, celestite and melilite.*

Miscellaneous.—Under the title "Rapid Qualitative Examination of Mineral Substances," Moses and Wells[30] publish a scheme for the detection of minerals. The blowpipe method is used with the metallic minerals, but in the silicate group a mixture of the dry and wet methods is made use of. From a hasty reading of the scheme it seems to be a practicable and convenient one.

[28]Neues. Jahrb., 1893, I, p. 49.
[29]Zeits. d. deutsch. geol. Ges., 1892, p. 224.
[30]School of Mines Quart., Nov., 1892, p. 25.

From The American Naturalist, April 1st, 1893.

The Petrography of the Abukuma Plateau, Japan.—The northern half of the Japanese Archean area, the Abukuma Plateau, is thought by Koto[2] to consist of a series of Laurentian granites and pressure gneisses, cut by younger granites and other eruptives, and overlying these a series of schists, divided by the author into lower and upper Huronian. The Laurentian granitic are an older amphibole-biotite variety and a younger, intruding biotite granite. The former contains, in addition to the usual granite components, microcline, and a bluish-green, weakly pleochroic hornblende, that very frequently plays the *role* of an ophitic groundmass for the other constituents. This granite passes by dynamo-metamorphism into foliated phases, in which the various minerals have been compressed, and the quartz, in addition, granulated. The Huronian (?) beds are principally schists and gneisses, that differ from the Laurentian gneisses in having the plane parallel structure, i. e., they are composed of bands of different composition. The most important schists of the lower division are: gneissic mica schists, containing andalusite and sillimanite, two-mica schists, one of whose constituents is margarite, garnet-biotite schist and hornblende schist. A peculiar member of the series is a titanite amphibole schist, consisting of bands whose structure is granular. Its black bands are made up of green-hornblende, plagioclase and a little biotite, and its white ones of sphene and granular sahlite in a groundmass of altered feldspar. The upper Huronian series embraces foliated amphibolites, mica-schists, and green schists that may be tufas. The distinction between the lower and upper members of the group seems to be based mainly upon petrographical characteristics. Among the rocks cutting these various schists may be mentioned an amphibole-picrite, pegmatites, and several varieties of diorite-porphyrite.

The Leucite-Tephrite of Hussak, from New Jersey.—The eleolite syenite eruption of Beemerville, N. J., was accompanied by basic extrusions now represented by the smaller dykes associated with the large eleolite-syenite dyke in this region. One of the most interesting of the basic dykes is the one at Hamburg, Sussex Co. It is from 15 to 20 feet wide and consists of a dark, tough, biotite rock, holding

[1]Edited by Dr. W. S. Bayley, Colby University, Waterville, Maine.
[2]Jour. Coll. Science, Imperial University, Japan, v, 8., p. 197.

spheroidal inclusions, that have been taken for Hussak[3] to be leucites. Kemp[4] has recently examined this rock very carefully, and now describes it as composed of biotite and pyroxene imbedded in an isotropic groundmass that is chiefly analcite. The biotite is dark brown and the pyroxene of a faint yellow color, with an extinction of 33°. The spheroidal inclusions are analcites, about whose ruins are often grouped grains of biotite and crystals of sphene. An analysis of one of the spheroids, after deducting 3.886 % of Ca Co$_3$, gave:

H$_2$O	SiO$_2$	Al$_2$O$_3$	Fe$_2$O$_3$	CaO	K$_2$O	Na$_2$O
6.31	52.44	26.44	.43	1.94	3.54	8.90

As to the origin of the analcite the author is not certain. It may have been derived either from leucite, in which case the rock would be a leucite-tephrite, as considered by Hussak, or it may be alteration product of nepheline.

A Sodalite-Syenite from Montana.—In the mountains forming the northern portion of Montana, Lindgren[5] and Melville have discovered post-cretaceous quartz-porphyrites, lamprophyres, augite-trachytes, analcite basalts, and a peculiar sodalite-syenite, somewhat resembling certain rocks described by Chrustschoff from Russia. The Montana syenite is from Square Butte, situated thirty miles southeast of Fort Benton. It is a light gray eruptive, associated with sheets of theralite and analcite basalt. Macroscopically it consists of lath-shaped feldspars, prisms of hornblende and pale brown grains of sodalite. In addition, analcite and plagioclase are discoverable under the microscope. Many of the feldspar crystals are corroded in an extraordinary manner and the cavities thus formed in them are filled with analcite which is believed to be an alteration product of albite. The hornblende is very dark brown, almost opaque, with a strong pleochroism, an extinction of 13° and a density of 3.437. Its analysis indicates its identity with the variety barkevikite:[6]

H$_2$O	SiO$_2$	Al$_2$O$_3$(TiO$_2$)	Fe$_2$O$_3$	FeO	NiO	MnO	CaO	MgO	Na$_2$O	K$_2$O
.24	38.41	17.65	3.75	21.75	tr	.15	10.52	2.54	2.95	1.95

The sodalite is quite fresh. It forms irregular grains that are bounded by crystal faces when in contact with analcite. It was evidently formed after the feldspar but before the analcite. The composition of

[3] AMERICAN NATURALIST., March, 1893, p. 274.
[4] Amer. Journ. Sci., Apr., 1893, p. 298.
[5] Amer. Journ. Sci., June, 1893, p. 286.
[6] AMERICAN NATURALIST, June, 1890, p. 576.

the rock as calculated from its analysis is : 23% hornblende, 50 orthoclase, 16% albite, 8% sodalite, and 3% analcite.

The Anorthosites of Canada.—The Canadian geologists have long considered the Laurentian of northern North America as consisting of an upper and a lower division, of which the latter rests unconformably upon the former. This upper division is made up largely of basic schists to which the name Novian was given by Hunt. Adams[1] has examined all of the important occurrences of the supposed schists, and has discovered that in all cases they show an irruptive contact with the surrounding gneisses, which they evidently cut. They are thus unquestionably post-Laurentian, and, from their relations to the overlying rocks, they are thought to be pre-Cambrian. The dark rocks are anorthosites—aggregates of plagioclase, with a little pyroxene, olivine and some accessories—which are in places schistose, and in other places are connected genetically with gabbros. The schistosity of the rock is accompanied by the possession of cataclastic structure, regarded by the author as due not to dynamic processes, but to the movement of the magma just before final consolidation. The plagioclase of the rock which is by far its most prominent component, is a labradorite so filled with tiny inclusions of microlites, thought by the author to be ilmenite tables, that fragments of the mineral are dark and often show the play of colors so beautifully seen in the labradorite of Labrador. The pyroxenes are a weakly pleochroic green augite, and a strongly pleochroic hypersthene. Hornblende, biotite, quartz, garnet and zircon are also present in small quantities in all specimens of the anorthosite. In the Saguenay river occurrence, olivine is enclosed in the plagioclase, and between it and the latter mineral is a reaction rim, composed of an inner zone of hypersthene, and an outer one of actinolite, including many small, green spinels. All the occurrences of the rock in Canada are briefly described, and with them are compared similar occurrences found elsewhere.

The Melibocus " Massiv " and its Dyke Rocks.—The peak of Melibocus[a] in the Odenwald consists mainly of a medium-grained white granite to the West, and a complex of schists and gneisses to the East. The granitic constituents, orthoclase and quartz, are usually aggregates of small grains variously orientated, and the biotite shows evidence of having been subjected to pressure. Near the contact with

[1] Neues Jahrb. f. Min., etc., B. B. viii, p. 419.

[a] C. Chelius: Notizbl. d. Ver. f. Erdk. z. Darmstadt, 1892, iv, F. 13 H., p. 1.

the surrounding rocks the granite becomes gneissic, and everywhere it is cut by dykes of aplites, porphyries and lamprophyres. Where the aptites penetrate the gneisses they possess the usual characteristics of these rocks, but where they pass from the schists into the granite they become porphyritic, showing a fine grained groundmass of quartz, orthoclase and mica and numerous phenocrysts of the same minerals and garnet. Like the granite the aplite components exhibit evidences of the effect of pressure. The large crystals are granulated and the rock's structure is more or less schistose. For this aplitic rock with porphyritic crystals the author, Chelius, used the name Alsbachite. An analysis of an alsbachite from the northwest side of the mountain gave:

SiO_2	Al_2O_3	Fe_2O_3	FeO	MnO	CaO	MgO	K_2O	Na_2O	H_2O
74.13	12.61	2.87	.86	.16	1.60	.23	2.13	4.55	.66

The dioritic aplites, malchite, luciite and orbite are also represented among these dyke rocks—the malchite being the panidiomorphic diorite aplite, the luciite the hypidiomorphic granular forms, and the orbite the corresponding porphyritic phases. One of the luciites is described as made up almost exclusively of plagioclase and hornblende. Among the lamprophyric dykes, mention is made of a gabbrophyre, or odinite, which differs from the gabbro-aplite, beerbachite, in consisting of phenocrysts of plagioclase and colorless augite in a matrix of plagioclase laths and hornblende needles, while the aplite is a panidiomorphic aggregate of diallage and feldspar, with the addition, sometimes, of hornblende crystals that enclose the other constituents. The descriptions of all these rare rocks are very brief.

The Granites of Argentina, S. A.—In an elaborate description of 185 hand specimens of stock and dyke granites and pegmatites, cutting the archean and paleozoic beds of Argentina, and of younger granites cutting these older ones, Romberg* discusses at some length the origin of the micropegmatitic intergrowth of quartz and feldspar. He believes that in the rocks studied by him the quartz in the intergrowths is secondary, and that it has originated in the decomposition of orthoclase and plagioclase. Many micro-photographs accompanying the author's article illustrate clearly the steps by which this conclusion was reached. Although the description of the specimens is exhaustive, it contains no points of special interest. At the conclusion of the paper is a list of the specimens examined, with their localities,

*Neues Jahrb. J. Min., etc. B. B. viii, p. 275.

and appended to it are twelve plates containing seventy-two micro-photographic reproductions of their sections.

New Minerals.—*Geikielite.*[10]—This mineral was found as pebbles in the gem washings near Rakwana, Ceylon. It is essentially a magnesium titanate, $MgTiO_3$, corresponding to the calcium compound, perofskite. The mineral is bluish-black and opaque, with a brilliant lustre, and possessing two cleavages at right angles to each other. Its density is 3.98 and hardness 6.5. In thin section it is translucent with a purplish-red tint, and in converged light it shows a uniaxial figure.

Baddeleyite,[10] also occurring as pebbles in the above-mentioned locality, is a black substance with a density of 6.02 and a hardness of 6.5, thus strongly resembling columbite. Under the microscope small fragments are seen to be dichroic in greenish-yellow and brown tints, and to possess a biaxial symmetry. The crystallization is thought to be monoclinic, though only a few plans could be detected on the specimen. In chemical composition the substance is zirconia ZrO_2.

Folgerite, blueite and *whartonite* are all nickel-iron-sulphides from the Sudbury nickel mines at Algoma, Ontario. Emmens[11] describes the first named as a massive, bronze-yellow substance, with a grayish-black streak, a density of 4.73 and hardness 3.5. Its composition (Fe $= 33.70$; Ni $= 35.20$; S $= 31.10$) corresponds to Ni Fe S_2.

The *blueite* is also massive. Its color 'is olive-gray or bronze; its streak black, density 4.2 and hardness 3–3.5. Its analysis yielded Fe $= 41.01$; Ni $= 3.70$; S $= 55.29$, corresponding to pyrite with a thirteenth of the Fe replaced by Ni. Unlike pyrite, however, it dissolves easily in nitric acid, without the precipitation of sulphur.

Whartonite differs from blueite in containing more Ni. Its composition is Fe $= 41.44$; Ni $= 6.27$; S $= 52.29$, corresponding to (Fe Ni) S_2 in which Fe:Ni $= 7:1$. Its hardness is 4, density 3.73, and color and streak like those of blueite.

Hauchecornite is another nickel mineral. It is described by Scheibe[12] from the Friederich mine in the Hamm mining district, Germany. It is found in bronze-yellow tetragonal crystals, with a hardness of 5, and a density of 6.4. It is thought to have the composition corresponding to the formula Ni (Bi. Sb. S), though analysis yields discordant results.

Cuprocassiterite was described by Ulke[13] from the Etta mine, South

[10] Fletcher: Nature, Oct. 27, 1892, p. 620.

[11] Jour. Amer. Chem. Soc., Vol. xiv, No. 7.

[12] Jahrb. A. preuss. geol. Landeranst, 1891, p. 91.

[13] Proc. Amer. Inst. Min. Engineers, Feb. 1892.

Dakota, but the author's data were so scanty that Headdon[14] has thought it advisable to add a small additional contribution to the literature of the mineral. This last-named writer obtained a small quantity of what he supposed to be Ulke's new mineral from both the Etta and Peerless mines, and found upon examination that in the interior of a small mass from the Peerless mine is a nucleus of stannite containing a little cadmium. Intergrown with this and also forming an envelope around it is a green clayey substance, which, upon its exterior, passes into a yellow earth. The green substance has a density of 3.312–3.374. Its analysis shows it to be a mixture of about 7 SnO_2, 6 CuO, 2 FeO and 11 H_2O. The author regards it is an alteration product of stannite, but not as a well-defined mineral species.

New Edition of Rosenbusch's Volume on Minerals.—The new edition of Professor Rosenbusch's[15] Microscopic Physiography of the Rock-forming Minerals is an enlargement rather than a revision of the second edition. There is no material difference in the arrangement of the matter in the two editions, but there have been large additions made in the later volume in the shape of descriptions of new petrographical apparatus and methods, and in the number of minerals treated. The plates illustrating the text have been decreased by one. The remainder are much better executed than was the case in the earlier volume.

Mineral Syntheses.—Michel[16] has obtained *melanite* garnets and *sphene* crystals by cooling slowly a mixture of 10 parts titanic iron, 10 parts calcium sulphide, 8 parts silica and 2 parts carbon, that had been heated to 1200° for five hours.

Crystallized *leucite, potassium cryolite* and *potassium nepheline* results from the fusion of silica or of fluosilicate of potassium and alumina with an excess of fluoride of potassium. Prolonged heating produces leucite, and potassium cryolite. Less prolonged treatment yields a potassium nepheline, which crystallizes in negative orthorhombic prisms.[17].

Instruments.—For measuring the curves of isotherms on mineral plates Jannetaz[18] has constructed a new ellipsometer, which it is

---

[14] Amer. Jour. Sci., Feb., 1893, p. 105.

[5] H. Rosenbusch: Mikroskopische Physiographie der petrographisch wichtigen Mineralien, Stuttgart, 1892, pp. 712, Fig. 239, etc.

[16] Comptes Rendus, Vol. cxv, p. 880.

[17] Duboin, Bull. Soc. Franc. d. Minn., Vol. xv, p. 191.

[18] Bull. Soc. Franc. d. Minn., Vol. xv, p. 237.

believed will enable its user to measure accurately the axes of the isothermal ellipses, and to determine rapidly in each case whether apparently circular isotherms are in reality circles or slightly eccentric ellipses.

A new machine for cutting and grinding thin sections of rocks and minerals, with stored electricity as the motive power, is described by G. H. Williams[19].

Rock Separations.—Thallium-silver-nitrate [$Tl\ Ag\ (NO_3)_2$], according to Retgers,[20] is an excellent medium for the separation of mineral grains of great density. The double salt fuses at 75°, and in the fused condition is clear and mobile. In this condition its specific gravity is 5, and this may easily be lowered by the addition of water. Its manipulation is simple. A small beaker containing the solid salt is placed in a water bath and heated. Upon its liquefaction the powder to be separated is added and the mixture is allowed to stand for a short time. As soon as a layer of clear liquid forms between the precipitated and the floating grains the beaker is plunged into cold water. The salt thus consolidates rapidly. The beaker is now broken and the heavy grains are collected by scraping and washing.

A new method of separating the constituents of rock powders, whose densities are above 2.60, has been devised by Dafert and Derby.[21] The principle involved is the suspension of small particles in gentle currents of water. The apparatus necessary for the operation is fully described by the authors. Separation is not complete between powders of nearly the same density, but there is a strong concentration of the heavier and the lighter ingredients in the two resulting portions of the separated material.

[19] Amer. Jour. Sci., Feb. 1893, p. 102.
[20] Neues Jahrb. f. Min., etc., 1893, Vol. I, p. 90.
[21] Proc. Roch. Acad. Sci., Vol. II, p. 122.

Reprinted from The American Naturalist, June 1st, 1893.

MINERALOGY AND PETROGRAPHY.[1]

Anorthosites and Diabases from the Minnesota Shore of Lake Superior.—Along them iddle stretch of the Minnesota shore of Lake Superior occur several exposures of a light-colored, coarse rock, consisting essentially of a basic plagioclase feldspar which, according to Lawson,[2] is sometimes bytownite; but more frequently anorthite or labradorite. This plagioclase is usually fresh and quite vitreous in appearance. It contains, as inclusions, small bleb-like masses of augite, plates and rods of the same mineral arranged parallel to the clinopinacoid, liquid enclosures, dust particles and small grains of hematite. In addition to the plagioclase there is also often present in the rock a small number of triangular augite plates between the feldspars. This rock which the author calls an " anorthosite," is found in knobs and bosses, and as boulders in the overlying Keweenawan eruptives. The rock is evidently an eruptive which is much older than the volcanic flows constituting a large proportion of the Keweenawan beds.

A second article by the same author[3] treats of the coarse diabase in "gabbro" sheets interpolated between the sedimentary beds of the Animikie. These are thought by the writer to be laccolitic in origin, i. e., to have been intruded between the sedimentaries after these had been solidified, and some of them even later than the time of deposition of the younger Keweenawan series. This conclusion is reached after a careful study of the contacts between the eruptives and the sedimentaries, which has brought to view the existence of contact phenomena at both the upper and lower surfaces of the diabase. The sheets of eruptives have been named the "Logan sills" in honor of Sir Wm. Logan, who was one of the pioneer geologists in the Lake Superior region.

The Volcanic Rocks of the Andes.—In a review of Küch's volcanic rocks of the Andes, Iddings[5] asserts that the chemical relations of the rocks studied indicate clearly that they all belong to the

[1] Edited by Dr. W. S. Bayley, Colby University, Waterville, Me.
[2] Geol. and Nat. Hist. Survey of Minn. Bull. No. 8.
[3] Ib. p. 24.
[4] Reiss and Stubel : Reisen in Sud-Amerika. Geologische Studien in der Republik Colombia, I. Petrógraphie, 1. Die Vulkanischen Gesteine bearbeitet von Richard Kuch, Berlin. 1892.
[5] Jour. of Geology. Vol. 1, p. 164.

same consanguinous group as do the Cordilleran rocks of Mexico and
the United States, and their nature indicates that the magma producing the Andes types has not yet become as highly differentiated as that
which yielded the corresponding volcanics in North America.

Basalts and Trachytes from Gough's Island.—Pirsson[6] has
examined some pebbles gathered from the beach of Gough's Island in
the South Atlantic. He finds two of them consisting of basalt, and the
others of trachyte glass and tuff. The glass is a pitchy-black mass,
filled with small pores and marked here and there by a phenocryst of
plagioclase. In thin sections it appears as a brown unaltered isotropic
substance containing magnetite, apatite, olivine and saidine phenocysts
and microlites of the last-named mineral. An analysis of the rock
gave:

SiO_2 TiO_2 Al_2O_3 Fe_2O_3 FeO MnO Mgo CaO Na_2O K_2O H_2O Total
61.22 .42 18.01 1.32 4.51 tr. .44 1.88 6.49 5.93 .46=100.68

Density = 2.210. The rock is thus shown to be unquestionably a
trachyte in spite of the fact that it contains occasional olivines. The
mineral evidently crystallized in an early stage of the rock's history, as
all its grains have been subjected to magmatic resorption.

The Origin of the Gneisses of Heidelberg.—In gneisses occurring in the region northwest of Heidelberg, Osann[7] finds lenticular
masses of graphitic and apatite schists, and therefore concludes that
the gneisses are of sedimentary origin. The rocks do not possess the
true gneissic foliation, since their feldspar, quartz, etc., do not show a
sequence in origin, nor do their micas exhibit the pressure phenomena
usually observed in the micas of other gneisses. Their structure is
described as the "hornfels structure" which is characteristic of contact
products. The graphitic schists consist principally of quartz, muscovite, graphite and flecks like the "Knoten" of contact rocks, which
are formed by the aggregation of plates of a green micaceous substance.
The apatite schist is composed of 55% apatite, 43% quartz, and 2% of
graphite, tourmaline and rutile. An analysis gave: CaO = 30.22 ;
P_2O_5 = 22.86 ; F = 2.16 ; Insol. = 43.52.

Petrographical News.—Retgers[8] communicates in a few brief
notes the results of his examination of rocks collected in southern

[6] Amer. Jour. Sci., XLV, 1893, p. 380.
[7] Mitth. gross. Bad. geol. Landesanst. Bd II, p. 372.
[8] Neues Jahrb. f. Min., etc., 1893, I, p. 39.

Borneo. Actinolite, smaragdite, and glaucophane schists are the most interesting foliated rocks studied. They contain, in addition to their characteristic components: epidote, garnet and orthoclase, most of which show the effects of torsion and pressure. A quartzite is remarkable in that it contains andalusite, sillimanite, rutile, zircon and tourmaline. The eruptives mentioned by the author as existing in this portion of the island are porphyrites, diorite, gabbro, peridotite, serpentine and a pyroxenite (augite-fels).

Analysis of cretaceous lithographic limestones from various localities in America and Germany give such discordant results that Volney[9] thinks it impossible to judge from analyses alone as to the commercial and technical value of such rocks. The organic matter in the stones contains nitrogen and traces of iodine. It is believed to be the residue of cretaceous fossils, and to be the cause of the peculiarly fine precipitation of the calcareous substance of good stones.

The term "poikilite" has already been referred to in this note as descriptive of a rock-structure produced by the inclusion of many differently orientated particles of some mineral irregularly distributed within large plates of another mineral. This structure has been described by so many petrographers as occurring in so many different rocks that Williams[10] suggests its general use and proposes "micropoicelitic" as the term descriptive of the structure when observed microscopically.

Some excellent examples of cone-in-cone structure in a concretion from the coal measures of Wolverhampton, England, are noted by Cole[11] as exhibiting clearly the crystalline structure of these bodies and their identity in mode of origin with spherulitic growths.

The rocks occurring at Cingolina in the Euganean Hills, described by Tchichatcheff[12] a few years ago, have been reinvestigated by Graeff and Brauns,[13] who find augite-syenite and olivine-diabase cut by dykes of hornblende and augite andesites. The plagioclase of the latter rock includes a large mass of the rock's groundmass which has crystallized largely as plagioclase with the same orientation as a thin zone of the same substance surrounding the corroded host.

New Minerals—*Sundtite.*—In some specimens of a silver ore from a mine at Oururo in Bolivia, Brögger[14] finds masses and crystals

[9] Journ. Amer. Chem. Soc , XIV, No. 10.
[10] Jour. of Geol., Vol. 1, p. 176.
[11] Miner. Magazine, X, p. 136.
[12] Neues Jahrb f. Min., etc., 1884, II, p. 140.
[13] Ib.. 1893, I, p 123.
[14] Zeits. F. Kryst, XXI, p. 193.

of a dark tetrahedrite-like mineral associated with stibnite and pyrite. The dark mineral is steel-gray, with a black streak. Its hardness is 3 –4 and density 5.5. Measurements of the crystals, some of which are 1 cm. long, indicated an orthorhombic symmetry. Twenty-one forms were observed, and from these the axial ratio $a : b : c = .6771 : 1 : .4458$ was determined. An analysis gave :

Cu	Ag	Fe	Sb	S		Total
1.49	11.81	6.58	45.03	35.89	=	100.80

which corresponds to the formula $(Ag_2 . Cu_2 . Fe) (SbS_3)_2$, or a salt of normal sulph-antimonic acid. Sundtite presents no analogies, either in composition or in its crystallographic characteristics, with other sulph-antimonates. Its 'nearest crystallographic relative is deschynite (RNb_2O_6). The new mineral is a commercially valuable ore of silver.

Melanostibian is another new mineral obtained by Igelström[15] from the celebrated manganese mine, Sjögrufvan, Grythyttan, Orebro, Sweden. It occurs as narrow veins in the dolomite, which is the bearer of all the ores of the mine. The mineral is in raven black, metallic-looking masses and tiny crystals, that are either tetragonal or orthorhombic. The streak of the mineral is cherry red, and its hardness 4. It is insoluble in dilute hydrochloric acid, but slowly dissolves in boiling acid. Its composition, corresponding to $6(Mn Fe)O Sb_2O_3$, was deduced from the following figures :

Sb_2O_3	FeO	MnO	CaO	MgO	H_2O	Total
37.50	27.30	29.62	1.97	1.03	1.06	= 98.48

Graphitite.—Upon treatment with nitric acid under certain conditions, the graphite from Ceylon, Norway and Canada yields an oxidation product that is different from the corresponding product obtained from the graphite of Fichtelgebirge, Siberia and Greenland. The materials of the two groups are therefore regarded by Luzi[16] as different, and as worthy of distinctive names. The mineral from the last-named localities is called graphitite.

American Minerals.—The *cookeite* of Paris and Hebron, Me., has been known for some time as a micaceous mineral closely related to the chlorites. In habit its plates are hexagonal, and are nearly always arranged in radial groups. These plates, according to Pen-

[15] Ib. p. 246.
[16] Ber. d. deutsch. chem. Ges., XXVI, p. 890.

field,[17] consist of an inner uniaxial hexagon surrounded by six segments extinguishing parallel to their edges and showing a biaxial interference figure. The mineral is monoclinic and is twinned like the clinochlor[18] from Texas, Pa. It is associated with quartz, lepidolite and tourmaline, and has probably been derived from the latter by alteration. An analysis of carefully selected material gave:

SiO$_2$ Al$_2$O$_3$ Fe$_2$O$_3$ CaO K$_2$O Na$_2$O Li$_2$O H$_2$O F   Total   O =F
34.00 45.06 .45   .04 .14 .19 4.02 14.96 .46=99.32—.19=99.13

This corresponds to the formula Li [Al(OH)$_2$]$_3$ (SiO$_4$)$_2$. The density of cookeite is 2.675.

The results of an examination of *zunyite* from the Charter Oak mine, at Red Mountain, Orange Co., Colo., and of *xenotime* from Cheyenne Mountain, El Paso Co., in the same State, have recently been communicated by Penfield.[19] The zunyite occurrence is five miles north of the original occurrence of the mineral first described by Hillebrand. The mineral is in little tetrahedrons scattered through an altered porphyrite. An analysis of these gave:

SiO$_2$ Al$_2$O$_3$ Fe$_2$O$_3$ Cl   F   H$_2$O P$_2$O$_5$ CaO Na$_2$O Total   O=Cl&F
24.1 157.20 .61 2.62 5.81 11.12 .64 .11 .48 =102.70—3.03=99.67

corresponding to [Al(Cl.F.OH)$_2$]$_6$ Al$_2$ (SiO$_4$)$_2$. Zunyite is found also in the mine as a white pulverulent mass resembling kaolin, but which consists of tiny octahedral crystals. The xenotime was from the tysonite locality described by Hidden[20] in 1885. It was a single fresh crystal of a brown color, implanted on a gangue of quartz and feldspar. Its density was found to be 5.106, and its composition : P$_2$O$_5$ = 32.11 ; Yt. Er = 67.78 ; Ign. = .18.

In the crystalline dolomite of Canaan, Ct., Hobbs[21] has discovered a rose-colored *talc*, that is noticeable for its large percentage of calcium and aluminium. Its analysis yielded:

SiO$_2$ Al$_2$O$_3$ MgO CaO FeO MnO H$_2$O Total
61.48 3.04 22.54 4.19 .77   tr. 5.54 = 100.56

The density of the mineral is 2.86. It is optically negative, and its axial angle 2 E is 15° 30′.

In a recent Bulletin of the United States Geological Survey, |Mel-

[17] Amer. Jour. Sci., XLV, 1893, p. 393.
[18] Ib. XLIV, p. 201.
[19] Ib. XLV, 1893, p. 396.
[20] Ib. XXIX, p. 249.

ville[21] has given the results of analysis of several American minerals as follows: *Natrolite* (I) from Magnet Cave, Ark.; a light-colored *tourmaline* (II) from Nevada Co., Cal.; *spessartite* (III) from Llano Co., Texas, and *bismuthinite* from Sinola, Mexico.   Figures follow:

	SiO₂	Al₂O₃	Fe₂O₃	FeO	MnO	CaO	MgO	K₂O	Na₂O	H₂O	F	B₂O₃

SiO_2 Al_2O_3 Fe_2O_3 FeO MnO CaO MgO K_2O Na_2O H_2O F B_2O_3
I   47.56 26.82   .20   .13 .09   15.40 9.63
II 36.40 33.64 3.13   1.51 10.01 .12 2.49 3.53 .74 8.74
III 35.93 18.08   4.60 31.77 8.48 .69 .17   .36

The spessartite contains also traces of TiO_2 and BaO.

Mineralogical News.—W. Ramsay[22] has discovered zonal growths of *epidote* substance in crystals of this mineral from the Sulzbachthal in Salzburg, Zöptan in Moravia, Arendal in Norway, Haddam, Conn., and from Traversella, Brosso and Ala in Piedmont.   The different zones possess not only different colors, but they have also different extinction planes and diffent refractive indices, as do also different portions of the same zones.

Fragments and small crystals of carbon with all the physical properties of *carbonado* have been prepared by Moisson[24] upon dissolving carbon in iron and cooling the mass slowly under pressure.

Miscellaneous.—McMahon[25] has elaborated a systematic course in micro-chemical analysis based on the production of the sulphates and double sulphates of the elements.   These salts are described as they appear on the object glass under the microscope, their habits are depicted and their constant peculiarities, if they possess any, are portrayed with some minuteness.   The methods of analysis developed by the author will prove of great convenience to petrographers if they are found as practicable as they are declared to be.

The eighth volume of the mineral resources of the United States, edited by Dr. D. T. Day,[26] contains statistical data for the calendar year 1891.   The total value of metallic products mined during this period amounts to over $181,000,000, and that of the non-metallic products over $241,000,000.   The most notable article in the volume is an historical description of the past "twenty years of progress in the manufacture of iron and steel in the United States."

[21] Ib. XLV, 1893. p. 404.
[22] Bull. No. 90. U. S Geol. Survey, p. 38.
[23] Neues. Jahrb. f. Min., etc., 1893, I, p. 111.
[24] Comptes Rendus, Feb. 6.   Ref. in Nature, Feb. 16, 1893, p. 370.
[25] Miner. Magazine, X, p. 79.
[26] Mineral Resources of the U. S.   Calendar year, 1891.   Washington. Govt. Printing Office, 1893.

Reprinted from The American Naturalist, October 1st, 1893.

MINERALOGY AND PETROGRAPHY.[1]

The Trachytes and Andesites of the Siebengebirge.—In the course of a discussion on the geological relations of the trachyte and andesite of the Siebengebirge,. Grosser[2] describes the various occurrences of these rocks and gives an outline of their petrographical characteristics. The trachytes he separates into typical, andesitic and aegerine varieties, and the andesites into trachytic and basaltic kinds. In the typical trachytes hornblende phenocrysts are frequent, but crystals of this mineral in the groundmass are unknown. Among the andesites the trachytic variety is noted for the absence of dark components from the groundmass and their rarity among the rock's phenocrysts. The basaltic andesite is rich in iron minerals, both as phenocrysts and as constituents of the groundmass. The order of eruption was trachyte, andesite, basalt.

A Variolitic Dyke in Ireland.—A variolitic dyke from Annalong, County Down, Ireland, resembles in the hand-specimen the variolites from Mt. Genévre. Cole[3] mentions it as consisting of devitrified glass, often containing skeleton crystals of magnetite, augite and plagioclase, and enclosing spherulites that are much larger toward the center than at the edge of the dyke. Thin selvages, 1 cm. in thickness, with very small spherulites scattered through them, exist on the sides of the dyke. Beyond these there is an abrupt transition to material containing the large spherulites. The selvages evidently cooled and lined the walls of the crevice now occudied by the dyke, before the interior filling consolidated; for not only is the transition between the substances of the two portions sharp, but the spherulites of the interior mass have in some cases grown from the line separating the two portions.

The Chemical Nature of Eruptive Rocks.—Lang[4] has returned to his study[5] of the chemical nature of eruptives. After a critical examination of many fresh specimens, the author concludes that the mineralogical nature of igneous rocks cannot be determined from

[1] Edited by Dr. W. S. Bayley, Colby University, Waterville, Me.
[2] Min. u. Petrog., Mitth., xiii, p. 39.
[3] Sci. Proc. Roy. Dub. Soc., 1892, p. 511.
[4] Min. u. Petrog. Mitth., xiii, p.115.
[5] Cf. AMERICAN NATURALIST, 1892, p. 334.

their chemical composition, but that types with the same general chemical relationships possess the same general mineralogical character. The author also gives his views on the relationships existing between the various rock types, as based on their calcium and alkali ratios, and, while not so stating it, he shows that the emanations from an eruptive center are consanguinous.

Norites in the Eastern United States.—Along a shear zone in the norite of Avalanch Lake in the Adirondacks, Kemp[6] finds what he believes to be a schistose phase of the rock in which several new minerals have been developed. The massive norite consists chiefly of plagioclase, with a little hornblende, enstatite and magnetite. In the schistose rock, which is much more basic than the norite, are broken pieces of plagioclase, shreds of hypersthene, grains of green monoclinic pyroxene, pink garnet, greenish-brown hornblende, biotite and magnetite, of which both the monoclinic pyroxene and the garnet are supposed to have been produced from the hypersthene and the plagioclase of the original norite. The schist resembles an eclogite. The same writer[7] records the discovery of a new occurrence of norite or of hypersthene gabbro at Artsdalen's quarry in Bucks County, Pa. It is associated with a limestone which is the matrix of a large number of metamorphic minerals. It is thought that this limestone may be a block brought from below by the eruptive. The region surrounding the quarry is underlain by pre-Cambrian rocks, but it is almost without exposures. The occurrence of norite here is interesting as affording a link connecting the otherwise separated Baltimore and Cortland areas of basic eruptives.

The Ottrelite Conglomerate of Vermont.—Reference has already been made in these notes to the discovery of an ottrelite conglomerate[8] in the Green Mountains of Vermont. Whittle[9] has now given us in more detail the description of its occurrence, and adds to this many items of interest concerning the dynamic schists associated with it. Among other things connected with the minerals of the conglomerate he mentions the secondary enlargement of clastic tourmaline grains and describes the alteration of microcline pebbles into quartz, sericite, biotite and albite. In one microcline there are many inclusions of limonite and rhombs of siderite. As the sericite grows it clears

[6] Amer. Journ. Sci., Aug., 1892, p. 109.
[7] Trans. N. Y. Acad. Sci., xii, p. 71.
[8] AMERICAN NATURALIST, April, 1894, p. 382.
[9] Bull. Geol. Soc. Amer., iv. p. 147.

the microcline of these, so that arouud each grain of the mica is a zone of pellucid feldspar, and on both sides of veins of the sericite are clear borders of microcline entirely free from inclusions of any kind.

Chalcedony and other Silicious Spherulites.—A well-illustrated article by Levy and Meunier-Chalmas[10] treats of various forms assumed by the molecule Si O_2 in the production of spherulites. Chalcedony has heretofore been regarded as a mixture of quartz and opal. The present authors have had an opportunity to study some excellent specimens of silica spherulites and concretions from the gypsum beds in the Paris Basin. Chalcedony and two new forms of silica, called by the authors quartzine and lutecite, are the components of these concretions. All three of these substances are fibrous forms of the same mineral, which is positive and biaxial, with an optical angle varying between 20°–35°. Thus they are different from quartz. The distinctions between the three varieties rest upon their habit. Chalcedony is elongated parallel to the base of the crystals, and quartzine parallel to the plane of their optical axis, while the lutecite fibers are elongated in a direction making an angle of 29° with the optical axial plane. The relation of the long axis of each variety to the optical constants of the mineral is carefully worked out, and the appearances of thin sections of their groupings are illustrated by eight beautifully executed photographs.

Petrographical News.—Andrea and Osann ascribe the existence of a porphyry breccia at Dorsenheim near Heidelberg to the crushing of porphyry by faulting and the cementing together of the fragments thus made by siliceous material.

A series of high dipping crystalline schists near Salida, Col., is regarded by Cross[12] as having originated by the alteration of great flows of basic and acid lavas erupted in Algonkian time. Though the rocks are now hornblende and micaceous schists, some of them still present a few of the structural features of diabases and porphyries.

Danalite from Redruth, Cornwall.—Tetrahedra of danalite at Redruth, Cornwall, are associated with quartz and arsenopyrite. Miers[13] mentions them as projecting from a layer of massive danalite with a thickness of from a quarter to half an inch. Some of the crys-

[10] Bull. Soc. Franc d. Min., xv. p. 159.
[11] Mitth. gross. Badisch. geol. Landesanst, ii, p. 365.
[12] Col. Sci. Soc., Jan. 2, 1893.
[13] Miner. Magazine, x, p. 10.

tals measure 30–50 mms. across. They are almandine-red in color, are translucent, and have a light pink streak, a hardness of 5.5 and a density of 3.350. An analysis gave:

SiO_2	FeO	MnO	ZnO	BeO	CaO	S	Total
29.48	37.53	11.23	4.87	14.17	tr	5.04	=102.62

corresponding to R.S. 7RO. 3SiO$_2$.

Mirabilite Changed to Thenardite.—Two crystals of mirabilite implanted on a mass of rock-salt from Aussee, Salzkammergut, that has been in the possession of the University of Vienna six years, have, in this time, so changed that they now consist simply of a thin shell composed of a crystalline aggregate whose inner surface is completely drusy. Within this crust there is usually a hollow, but occasionally a part of the hollow may be filled by a group of crystals like those forming the shell. These crystals are determined by Pelikan[14] to be *thenardites* of a short pyramidal habit, bounded by the planes P. ½P, P∞, ⅓P∞ and ∞ ∞, with an axial ratio of $a : b : c = .5970 : 1 : 1.2541$. The crystals had been kept during the six years in an air-tight enclosure at a nearly uniform temperature, so that the change from their original condition must have been due solely to the influence of the small amount of moisture within the enclosure.

Mineralogical News.—Crystals of the rare *uranatile* from Schneeberg, Saxony, and from the Joachimsthal, Bohemia, have been measured by Pjatnitzky,[15] who concludes that they are triclinic and not orthorhombic as Zepharovich supposed. Their axial ratio $a : b : c = .6257 : 1 : .5943$. The mineral has a citron or sulphur-yellow color, with very weak dichroism. *Uranophane,* according to the author should not yet be considered a species. Its chemical composition is the same as that of uranatile, but its crystallization has not yet been determined.

The rare plane 2O∞ has been detected by Pelikan[16] on salt crystals from Stannia, Galicia. Upon examining sections of *halite* from this locality, the author discovered in them many inclusions of petroleum zonally arranged. The cavities in which the oil is contained are either pear-shaped or are negative crystals, entirely or only partially filled with the liquid, which must have been under greater pressure at the

[14] Min. u. Petrog. Mitth., 1892, xii, p. 476.
[15] Zeits. f. Kryst., xxi, 1892, p. 74.
[16] Min. u. Petrog. Mitth., xii, p. 483.

time of its imprisonment. From the distribution of these inclusions the author concludes that the crystals were first cubes, the tetrahexahedra (20∞), and finally cubes, as at present.

Jannetaz[17] has made an analysis of the black garnet *pyreneite*, now the subject of so much discussion[18] in Europe, and has found it to consist of:

SiO_2	Al_2O_3	FeO	MgO	CaO	Total
39.4	10.0	18.6	1.0	31.21	=100.21.

It is thus neither melanite nor grossularite, but is intermediate in composition between the two. Its density is 3.7.

Miers[19] has succeeded in obtaining some excellent though tiny crystals of *orpiment* by dissolving in hydrochloric acid the marl in which nodules of this substance are found at Tajowa, Hungary. Under the microscope the little crystals appear with the orthorhombic symmetry. oP is the plane of their optical axes. Their axial angle for sodium light is 70° 24' in air.

The same ¡mineralogist[20] has repeated Gmelius' analaysis of *helvite* from Schwarzenberg, and has obtained this result:

SiO_2	FeO	MnO	BeO	Al_2O_3	CoO	S	Total
31.85	4.26	42.47	14.25	·74	3.16	4.81	=101.54

Dumortierite is recorded by Gonnard[21] as occurring in the feldspar of a granite vein cutting the gneiss in a quarry at Ternières, Francheville, Dept. of the Rhone, France.

The same writer[22] figures a few new types of *natrolite* crystals from the Puy-de-Dôm, and describes[23] the occurrence of crystals of *analcite* in the fissures of the porphyry at Agay, Canton Hyères, France.

Brazilite, analyzed by Blomstrand,[24] has the following composition:

ZrO_2	SiO_2	Al_2O_3	Fe_2O_3	CaO	MgO	Alk	Loss	Total
96.52	.70	.43	.41	.55	.10	.42	.39	=99.52

Experiments in Crystallization.—Hundt[25] has repeated Vogel-

---

[17] Bull. d. l. Soc. Franc. d. Min., xv, p. 127.

[18] AMERICAN NATURALIST, Oct., 1892, p. 849. Ib., Apr., 1893, p. 385.

[19] Miner. Magazine, x, p. 24.

[20] Ib., x, p. 10.

[21] Bull. Soc. Franc. d. Min., xv, p. 230.

[22] Ib., p. 221.

[23] Ib., p. 231.

[24] Neues Jahrb. f. Min., etc., 1893, I, p. 89.

sang's experiments on the crystallization of sulphur from its solution in carbon bisulphide thickened with balsam, and has discovered thereby some new facts regarding the phenomena connected with the formation of crystals. He finds the globulites aggregating into *liquid* spherules of sulphur that may remain liquid for several days. Grains of sulphur that are melted on a glass plate may also remain in a liquid condition for a long time—in some instances, three months—before they solidify. Upon agitation with the point of a needle they immediately become solid. The author declares that there is no tendency among the globulites to arrange themselves into definite groups, as Vogelsang reported to be the case. In the largest drops, however, they may take definite positions, whereupon the entire drop may be made to crystallize by shaking or agitating with a needle point. The formation of crystallites is contemporaneous with that of the globulites, the latter giving rise to the large drops, which, upon soldifying, become spherulites, and the former growing into microlites by the accretion of *invisible* particles. The crystallites do not grow by the addition of globulites. These bodies add themselves to the large drops, and never to the small, solid embryo crystals.

Miscellaneous.—A couple of *slags* from the lead ovens of Raibl, Austria, have been examined chemically by Heberdey.[25] The composition of different portions of the various specimens were carefully worked out. In one specimen crystals of a lead-zinc *olivine* were found, the analysis of which yielded:

SiO_2	PbO	ZnO	MgO	FeO	CaO	Total
16.62	61.50	18.16	1.99	1.69	tr =	99.96

Their density is 5.214 and axial ratio $a : b = .8592 : 1$. In an appendix to his main article the author gives the results of analyses of the limestone in which the galena smelted in the furnace occurs. One of these analysis yielded: $CaCO_3 = 53.50$; $MgCO_3 = 46.51$; Fe, Tl, Li = traces.

Dunnington and Whitlock[27] communicated the results of an analysis of a *black soil* from a point in the valley of the Red River of the North, about fifteen miles south of Winnipeg, Manitoba, and Corse and Baskerville[28] the results of analyses of *glauconite* sand from near Han-

[25] Mitth. d. miner. Inst. d. Univ. Kiel. B 1. H. 4., p. 310.
[26] Zeits. f. Kryst., xxi, 1892, p. 56.
[27] American Chem. Journal, 14, 1892, p. 621.
[28] Ib., p. 627.

over Court House, Virginia. Analyses follow (I, black soil; II, glauconite):

Sand	SiO_2	TiO_2	Al_2O_3	Fe_2O_3	CaO	MgO	SO_3	CO_2	P_2O_5	K_2O	Org.	H_2O
I 59.82	5.45	.64	7.14	4.00	.61	.61	.03	.37	.13	1.91	12.49	6.86

Quartz	SiO_2	Al_2O_3	Fe_2O_3	FeO	CaO	MgO	K_2O	Na_2O	H_2O		Total
II 2.76	47.45	7.33	12.03	9.43	.57	2.90	5.75	.42	9.85	=	98.49
8.22	43.34	6.62	15.16	8.33	.62	.95	4.15	1.84	10.32	=	99.55

Schwartz has treated in a comprehensive essay[29] the history of the observations on *reciprocal changes* produced in polymorphous bodies under different conditions of temperature, and has, in addition, given the results of some independent observations of his own. The substances that have been experimented upon are: AgI, KNO_3, NH_4, NO_3, $AgNO_3$, $Rb (NO_3)$, boracite, perchlorethane, tetrabrommethane, and copper, nickel, zinc and cobalt, sodium-uranyl acetates.

Ch. Friedell[30] has examined carefully a specimen of the meteoric iron from Cañon Diablo, Arizona, and, as a result of his study, has concluded that particles of black diamond (carbonado) are disseminated through its mass. A combustion of the residue obtained upon treatment of the iron by acids leaves no doubt but that the material consists principally of carbon.

[29] Gekronte Preisschr. Univ. Goëtinger, 1892.
[30] Bull. Soc. Franc. d. Min., xv, p. 258.

Reprinted from The American Naturalist, November 1st, 1893.

MINERALOGY AND PETROGRAPHY.[1]

The Schists of Southern Berkshire, Massachusetts.— The sericite schists of southern Berkshire Co., Massachusetts, and northern Litchfield Co., Conn., contain phenocrysts of feldspar, garnet, staurolite, tourmaline, biotite, and ottrelite, imbedded in an aggregate of feldspar, quartz and sericite, which contains, besides the phenocrysts, a large number of metamorphic minerals. The large feldspars are often filled with secondary granophyre, and this mineral, the garnet and the tourmaline, are frequently built out by secondary enlargements. The core of the feldspar is so often bounded by crystal outlines that Hobbs[2] regards the mineral as having resulted from the re-crystallization of the clastic grains of the original rock. The garnets, in addition to their peripheral enlargements, are often possessed of a rim of staurolite and magnetite crystals, supposed to be the product of reactionary action between the garnet and the surrounding minerals. The author believes the phenocrysts to have been developed by static metamorphism (simple pressing) from the constituents of a fragmental rock.

The Phonolytes of the Hegau.—The phonolytic rocks of the Hegau, Eifel, Germany, so well-known because of the beauty of their hauyne constituents, have been subjected to a comparative study by Cushing and Weinschenk,[3] who find them not all phonolites, as they have heretofore been regarded. The essential characteristic constituents of the group are sanidine, nosean, hauyne, nepheline, leucite, augite and aegerine, and the accessories, biotite, apatite and zircon. All the rocks are more or less porphyritic, with sanidine and the members of the hauyne group in two generations. Of the latter the larger crystals and those of the first generation are hauyne; the smaller, those of the second generation, nosean. The former are always more or less altered into zeolites, while the latter are usually fresh. Contrary to the general statement made with regard to these two minerals, the hauyne is not always blue nor the nosean colorless, but rather is the opposite the case. An important discovery made during the investigation is to the effect that nepheline is by no means common

[1] Edited by Dr. W. S. Bayley, Colby University, Waterville, Me.
[2] Bull. Geol. Soc. Amer., Vol. IV, p. 167.
[3] Minn. u. Petrog. Mitth., XIII, p. 18.

in the Hegau rocks. In a few of them the mineral is abundant in the usual form. In others it is only sparingly present, while in still others it is absent so far as could be learned. Consequently, the rocks fall into several classes. The Hohentwiel occurrences are of nosean-phonolites, in which nosean is abundant in the groundmass and nepheline absent. The specimens from Mägdeberg and Schwindel are nosean-ophyres (corresponding to the leucitophyres), in which nosean and nepheline are both present. At Staufen, two types were found, one a leucite-phonolite, and the other a true phonolite (nepheline-phonolite). The rock of Gonnersbohl is a hauyne-bearing trachyte or a trachytic phonolite. Each of these types is briefly described, and at the conclusion of the paper a few pages are devoted to an account of the tufa associated with them.

The Rock of a New Island, off Pantelleria.—An island, measuring one kilo. in length, and two hundred metres in width, was projected above the water off Pantelleria during the earthquake week beginning Oct. 14, 1891. The new island is an aggregate of loose blocks and solid lava, whose characteristics have been described by Foerstner.[4] The material in his possession was mainly a black pumi-

[4] Minn. u. Petrog. Mitth., XII, 1892, p. 510.

ceous basalt of the composition:

SiO$_2$ TiO$_2$ Al$_2$O$_3$ Fe$_2$O$_3$ Feo MnO CaO MgO K$_2$O Na$_2$O H$_2$O   Total
44.64 5.86 12.74 4.21 11.17 .20 10.12 5.82 1.41 4.31 .51=100.99

Attention is called to the large quantity of TiO$_2$ revealed by the analysis. Under the microscope the groundmass of the rock is seen to be a dark glass filled with highly colored microlites, and enclosing phenocysts of anorthite, olive-green augite, olivine and magnetite. The glass is sometimes in large quantity and at other times is present only in traces. The rock is a tachylitic basalt like that of Pantelleria and the other neighboring islands.

Petrographical News.—The pyroxenite of Duerne, Dept. of the Rhone, France, is an aggregate of orthoclase, pyroxene, oligoclase, garnet, and quartz. The structure of the rock varies from pegmatitic to granular. Its pyroxenic component is described by Gonnard[5] as light green in color, and as often possessing crystal outlines. It includes within its mass many crystals of sphene. Druses of vesuvianite line the walls of crevices in the rock, and galena is not an uncommon constituent of tiny veins traversing it.

[5] Bull. Soc. Franc d. Min., XV, p. 232.

In the northern Hardt Mountains, near Obbersweiler, Waldhambach and the neighboring regions, are biotite and hornblende gneisses that are probably squeezed granites, schists and graywackes, altered by an intrusive biotite granite and cut by other granites, a kersantite dyke cutting the intruded granite, and a sheet of quartz melaphyre overlying all these as a lava flow, the whole comprising the mass of the mountains. All these rocks Leppla[6] discusses in a recent article, describing the melaphyre as consisting of a groundmass of plagioclase and quartz enclosing phenocrysts of feldspar, red olivine pseudomorphs, quartz and bastite. The quartzes are all surrounded by aureoles of augite, just as are the quartz inclusions in many basic rocks.

In a monograph in the Kaiserstuhl in Baden, Knop[7] gives a general view of the geology, mineralogy and chemistry of this interesting volcanic region, in addition to statements concerning its hydrography, botany, history, etc. All the minerals known to the region are described at considerable length, and over a hundred and fifty pages of the book are devoted to descriptions of its interesting rocks, phonolites, andesites, tephrites, basanites, basalts, limburgites among the volcanics, and several others of sedimentary origin. The author treats the hill as an old volcano, and attempts to explain the variety in its products upon the Bunsen theory of mixed magmas.

A two-mica gneiss[8] constitutes the principal rock of the Valley of Miñor, Provice Pontevedra, Spain. On the peninsula of Santa Marta it is cut by a diabase with faintly pleochroic augite. At Montè Galeñeiro the micaceous gneiss is replaced by a hornblendic variety in which the prominent amphiboles are glaucophane and a green variety opaque to light vibrating parallel to *c*.

Chelius[9] describes very briefly several occurrences of nepheline basalt from the Odenwald, Germany, and records the analyses of the red gneiss of Steinkopf, of the dark biotite gneiss of Bockenrod, of basalt from the Häsengebirg near Urberach, of granite from the .Melibocus massiv, and the results of silica and specific gravity determinations of many other rocks from the same region, among which may be mentioned malchite and alsbachite.

[6] Zeits. d. deutsch. geol. Ges., XLIV, p. 400.
[7] Der Kaiserstuhl in Breisgau. Ein naturwissenschaftliche Studie von Dr. A. Knop. Leipzig. W. Engelmann, 1892, p. 538 and fig. 89.
[8] Quiroga: Actas d. l. Soc. Esp. d. Hist. Nat., XXI, 1892, pp. 4 and 8.
[9] Notizbl. d. Ver. f. Erdk. Darmstadt., IV, 1891, H. 12.

Though the parallel growths of augite and hornblende, with the latter mineral surrounding the former, are common, the reversed phenomenon is rare. Hobbs,[10] however, has recently pictured an example of light green amphibole completely encircled by colorless augite from an augite-hornblende rock occurring at New Marlboro, Mass.

Analyses of American Minerals.—Several analyses of dodecahedral crystals of *aguilarite* from Guanajuato, Mexico, have been made by Genth and Penfield.[11] That from the purest material gave: Ag = 84.40 % ; Cu = .49 % ; S = 11.36 % ; Se (diff) = 3.75 %. The mineral is thus an argentite with an eighth of its S replaced by Se. *Metacinnabarite* particles disseminated through barite from San Joaquin, Orange Co., Cal., gave the same authors: Hg = 85.89 % ; S = 13.69 % ; Cl = .32 %. The mineral supposed to be *leucopyrite*,[12] from Alexander Co., N. C., is *lollingite*, whose composition is Fe = 70.83 ; Cu = tr; As = 27.93 % ; S = 77 %. *Rutile* crystals with the habit of cassiterite are found in the quartz decomposition products of the orthoclase from West Cheyenne Cañon, El Paso Co., Col. They are iron black with a density of 4.249, and the composition: SnO_2 = 1.40 ; TiO_2 = 91.96 ; Fe_2O_3 = 6.68. The *quartz* decomposition products referred to yield, upon analysis:

SiO_2	Al_2O_3	Fe_2O_3	Na_2O	K_2O	Loss	Total
96.63	.93	.85	tr.	.46	.95	= 99.82

Pieces of a large *danalite* crystal from the same locality, give:

SiO_2	BeO	CuO	ZnO	FeO	MnO	S.	Loss	Total—O
30.26	12.70	.30	46.20	6.81	1.22	5.49	.21	100.41

The danalite is associated with quartz, astrophyllite and a new yttrium calcium fluoride with a hardness of 4 and a density of 4.316. Its composition is: CaO = 19.41 ; $(Yt Er)_2O_3$ = 47.58, etc. The other minerals whose analyses are recorded by the authors, are: altered zircon (*cyrtolite*) from Mt. Antero, Col. ; *lepidolite*, from Tanagama Yama, Japan, and *fuchsite*, from Habershaw Co., Ga. The analyses of the lepidolite and fuchsite follow:

[10] Sci., Dec. 23, 1892, p. 354.
[11] Amer. Jour. Sci., Nov., 1892, p. 881.
[12] Bull U. S. Geol. Survey, No. 74, p. 26.

	SiO_2	Al_2O_3	F_2O_3	MnO	MgO	CaO	Li_2O	Na_2O	K_2O	H_2O	Fl	Cr_2O_3	CuO
L.	53.84	17.76	3.25	2.77	.05	.37	4.60	1.55	10.90	.65	7.78		
F.	46.73	29.00	2.59		3.03			.26	9.25	6.04		2.73	.14

Among some analyses[13] made in the laboratory of the University of Virginia are the following, which are of interest to mineralogists: *Cuproplumbite* from Butte City, Montana, analysed by De Bell, gave: Cu = 61.32; Pb = 18.97; S = 17.77; quartz = 1.58; corresponding to 5 Cu_2 S. Pb S. *Calamine,* from New River, Wythe Co., Va., yielded Jones: SiO_2 = 25.33; ZnO = 67.15; H_2O = 7.47; Total = 99.95. *Parantite,* from a pocket in a corundum vein of the Hiawassee Corundum mine, Hayesville, Clay Co., N. C., is associated with decomposed albite and various chlorites. It is in rounded blue-gray lumps, having a density of 2.75. Analysed by Berkeley, it gave:

SiO_2	Al_2O_3	CaO	Na_2O	H_2O	Total
47.54	34.03	17.23	1.82	1.02 =	101.64

North American Minerals.—Remarkably large crystals of *selenite* have been found by Talmage[14] in the drainage area of one of the side cañons of the Tremont River, Wayne Co., Utah. Gypsum in seams cuts through the sandstone and argillite of the region in great profusion. The largest crystals of the minerals were in a geode-like cave, left exposed as a hollow mound in the slope of a hill. The interior of the cave was studded with great columns and slabs, extending from its sides sometimes to a distance of 51 inches. Many of the crystals are transparent throughout their entire length.

Fairbanks[15] describes the *rubellite* and *lepidolite* of southern California as occurring in a pegmatite vein cutting norite near Pala, west of Smith's Mt, San Diego Co. Besides the feldspar and the quartz there are associated with the two minerals above mentioned: muscovite, hematite, and green and black tourmalines.

A few very fine *datholite* crytals from the Lacy Mine, Loughboro, Ontario, have been measured by Pirsson.[16] The manner of their occurrence is not certainly known, but they appear to be in a vein penetrating an eruptive rock. The crystals are described as the finest yet

[13] Amer. Chem. Jour., Vol. XIV, 1892, p. 620.
[14] Science, XXI, 1893, p. 85.
[15] Ib. XXI, p. 85.
[16] Amer. Jour. Sci., Feb., 1893, p. 100.

found in America. They are transparent, yellowish, and in size the largest measure 3 x 2.5 x 2 cm. Their habit is prismatic parallel to *a*, and each individual is bounded by many faces.

In and upon calcite crystals lining some of the geodes of Keokuk, Iowa, Keyes[17] announces the discovery of very handsome tufts and radiating masses of *millerite*.

Beds of specular *hematite*, intermixed with *martite*, are reported by Hill[18] as abundant at the junction of diorite and limestone in many localities within the State of Coahuila, Mexico.

Physical Properties of Minerals.—A series of new determinations of the specific heat of *boracite* at different temperatures is reported by Kroeker.[19] The materials experimented upon were four transparent crystals from Linneberg, one piece being from a large crystal with a cubical habit, and the others fragments of dodecahedral crystals. In all cases it was found that the specific heat of the mineral varies with the temperature, and that the increment of variation increases rapidly between 250°–270°. Below 270° the cubic and the dodecahedral crystal gave similar results, above this temperature the results are different. For details of the experiments the reader must be referred to the author's paper.

Two articles of interest to mathematically inclined physical mineralogists are the one by Pockels[20] on the changes effected in the optical characterists of *alum* and *beryl* by pressure acting in a single direction, and the other, by the same author,[21] on the elastic deformation of piezo-electrical crystals in the electrical field.

Traube[22] finds that the following compounds, all of which form dextro-rotatory solutions, are hemi-morphic, viz.: $Sr(SbO)_2(C_4H_4O_6)_2$, $Pb(SbO)_2(C_4H_4O_6)_2$, and $Ba(SbO)_2(C_4H_4O_6)_2 + H_2O$. All are also tetartohedral—the strontium and lead compounds being the first examples of hemimorphic tetartohedral substances crystallizing in the hexagonal system, and the barium salt the first instance among tetragonal bodies.

[17] Amer. Geologist, XI, p. 126.
[18] Amer. Jour. Sci. XLV, p. 111.
[19] Neues Jahrb. f. Min., etc., 1892, p. 125.
[20] Neues Jahrb. f. Min., etc., B. B. VIII, p. 217.
[21] Ib. B. B., VIII, p. 407.
[22] Ib. B. B., VIII, p. 269.

One of the micas[22] in the Mte. Dorè trachyte is an *anomite* with an optical angle of 41°.   When treated with boiling hydrochloric acid it loses its greenish color, with the extraction of its iron and magnesium, and becomes less strongly doubly refracting.   After an hour's treatment it becomes colorless and uniaxial, when its optical sign is negative.

Having examined seventy-one uniaxial minerals with respect to their heat conductivity, Jannetaz[24] finds that only five contradict his law that the major axis of the isothermal ellipsoid is parallel to the direction of the principal cleavage, and the minor axis normal thereto.

Instruments.—Laspeyres[25] describes a modification of the setting of the condenser above the polarizer of the microscope, that enables the observer to change rapidly from converged to parallel light, even when an object is being examined.   The lower nicol is in its usual position. The condenser is imbedded in a metal strip set into the stage, and sliding easily in a groove prepared for it.

A goniometer with two circles, enabling the operator to measure nearly all the planes on a crystal with one adjustment of the latter, is explained in detail by Goldschmidt,[26] who also illustrates its use by several examples.

[22] Bull. Soc. Franc d. Min., XV, 1892, p. 97.

[24] Ib. XV, p. 133.

[25] Zeits. J. Kryst., XXI, p. 256.

[26] Ib., p. 210.

Reprinted from The American Naturalist, December 1st, 1893.

PARTIAL INDEX OF SUBJECTS.

INDEX OF AUTHORS.

No. 13.

A

SUMMARY OF PROGRESS

IN

Mineralogy and Petrography

IN

1894

Petrography and Mineralogy

BY

W. S. BAYLEY.

Mineralogy

BY

W. H. HOBBS.

FROM MONTHLY NOTES IN THE "AMERICAN NATURALIST."

PRICE 50 CENTS.

WATERVILLE, ME.:
PRINTED AT THE MAIL OFFICE.
1895.

MINERALOGY AND PETROGRAPHY.[1]

The Granite of Santa Lucia, California, and a New Rock Variety Carmeloite.—The Santa Lucia Mountains[2] in the vicinity of Carmelo Bay, California, consist largely of a porphyritic granite whose phenocrysts of glassy orthoclase are corroded with inclusions of cloudy orthoclase, plagioclase, quartz, biotite, apatite and muscovite, which substances also constitute the groundmass of the rock. The striking feature of the inclusions is that their different areas are not only uniformly orientated with respect to each other, but they are also definitely orientated with reference to their host. They lie in certain definite planes within the phenocrysts, and their crystallographic axes are definitely arranged with respect to the axes of their hosts. The quartzes all lie with their vertical axes nearly perpendicular to the basal plane of the orthoclase, consequently in sections of the phenocrysts cut parallel to the basal pinacoid every included quartz grain exhibits the axial figure. Another feature worthy of notice is the tendency of the inclusions to idiomorphic forms, whereas, the same minerals in the rock's groundmass are always allotriomorphic. The facts of the idiomorphism of the inclusions and their definite orientation suggest to Lawson that these and their hosts are of contemporaneous age. This view is strengthened by the observation that many inclusions on the edges of the phenocysts have grown out into the surrounding matrix, in which, as has already been noted, the components are the same as those occurring as inclusions, but are much larger than these, and are allotriomorphically developed. This granite is cut by dykes of fine-grained aplite.

The rock to which the author has given the name Carmeloite, is a young volcanic, marked by all the characters of a recent lava. It is probably younger than the Monterey series of the Miocene, and older than the newer terrace formations of the region. Under the microscope the rock is seen to consist of phenocrysts of iddingsite, plagioclase and often augite in a matrix composed of a felt of small, lath-shaped plagioclase and granules of magnetite and pyroxene, lying in a glass containing numerous feebly polarizing globulites. There are six areas of the rock in the Carmelo Bay district, the occurrences differing mainly in the quantity of glass present, the presence or absence of

[1] Edited by Dr. W. S. Bayley, Colby University, Waterville, Me.
[2] A. C. Lawson, Bull. Dept. Geol. Univ. of Cal., Vol. I, p. 1.

iddingsite in the groundmass, and of augite among the phenocrysts. The occurrences differ also in their chemical composition, their silica contents varying between 52.83 % and 60.00 %. The analysis of one specimen (Sp. Gr. = 2.51-–2.54).

SiO₂ Al₂O₃ Fe₂O₃ FeO MnO CaO MgO K₂O Na₂O Ign.   Total
60.00 19.01 3.20 .68 tr. 4.10 1.28 2.79 6.97 4.30 = 102.33

Since the rock contains too much SiO_2 for a basalt, and too little for andesite, and because of the prominence of iddingsite as one of its essential components, the author prefers the new name, Carmeloite, to any already in use among petrographers.

The Ancient Rocks of Southern Finland.—In the German resumé of his article on the old rocks of southern Finland, Sederholm[3] divides these into two groups—the Archean and the Algonkian, and the first of these groups into two sub-groups. The older Archean consists of phyllites, gneisses, micaceous and other schists and granular limestone, cut by granite and diorite. All the members of the series have been subjected to dynamic metamorphism on an enormous scale. The schists are supposed to have originated both in sedimentary and in erruptive rocks. The younger Archean schists are phyllites, micaschists, sandstone-schists, and a greenstone schist that was originally a uralite porphyrite occurring as surface flow. These are cut by a red granite that is sometimes porphyritic and often pegmatitic. It shows no evidence of having been subject to great pressure, but nevertheless it is foliated—a consequence, according to the author, of flowage. The Algonkian rocks are all fragmental, and above them are the Rapakivi granite and a diabase, both of which are effusive. A younger olivine diabase and a panidiomorphic gabbro are also thought to be volcanic flows.

Petrographical News.—Smith[4] has discovered that the supposed peridotite[5] of Manheim, N. Y., is an alnoite in which there is no pyroxene. It contains a large quantity of melilite in the typical forms, but the mineral is positive in the character of its double refraction, like the artificial melilite made by Vogt. Incidentally the author mentions that positive melilite exists also in the nepheline basalt of Wartenburg, Bohemia, and in the alnoite from Alno, Sweden.

[3] Fennia, 8, No. 3, p. 138.
[4] Amer. Journ. Sci., XLVI, p. 105.
[5] Cp. AMERICAN NATURALIST, Sept., 1892, p. 769.

About 600 miles north of the Falkland Islands in the South Atlantic, a fall of volcanic dust occurred on May 26, 1892.   Palache,[6] who has examined some of the material, finds it to consist of fragments of glass and pieces and crystals of orthoclase, plagioclase, green hornblende and magnetite, with a very small quantity of what appears to be pyroxene.   The character of the dust is thus andesitic.

New Minerals..—*Iddingsite* has been known for some time as a component of certain eruptive rocks from the far west, but not until Lawson[7] discovered it in the carmeloite of California, had its characteristics been carefully enough investigated to warrant its receiving a name.   As described by Lawson, iddingsite occurs as a phenocryst with well-defined crystal outlines.   It is of a bronzy color, has a very perfect cleavage and a hardness of 2.5.   Its cleavage lamellae are brittle. Before the blow-pipe the mineral is infusible, though it loses water when heated.   It is decomposed by acids after long treatment, but loses only its dark pigment, without alteration of its optical properties, when gently heated with hydrochloric acid.   Maximum density $= 2.839$. Its crystals possess in thin section the habit of olivine.   If the cleavage is regarded as pinacoidal, the other crystallographic faces are the prism, with a prismatic angle of about 80°, and another pinacoid, both of which are perpendicular to the cleavage.   The elongation of the crystals is in the direction of the second pinacoid.   If the cleavage is regarded as parallel to the macropinacoid,. b is in the cleavage plane, a is at right angles to it, and c is parallel to the elongation of the crystals.   The plane of the optical axes is the brachypinacoid, and the mineral is orthorhombic and negative; $a = A$, $b = B$ and $c = C$.   In thin section the color varies between yellowish green and chestnut brown, and the absorption is strong parallel to c.   The absorption formula is $C > B > A$.   The mean index of its fraction is low, and the double refraction strong.   Qualitative tests showed the presence of silicon, iron, calcium, magnesium, sodium and water.   In spite of the resemblance of its crystals to those of olivine, the author regards it as most probably an original separation from the magma that yielded the carmeloite.

Mackintoshite is the name given by Hidden and Hillebrand[8] to the original material from which the alteration product *thorogummite*[9] is derived.   Only a very small quantity was available for study.   This is

[6] Amer. Geol., June, 1893, XI, p. 422.
[7] Bull. Dept. Geol. Univ. of Cal., Vol. 1, p. 31.
[8] Amer. Jour. Sci.. XLIV, 1890, p. 98.
[9] AMERICAN NATURALIST, Jan., 1893, p. 72.

described as opaque and black. Its hardness is 5.5 and density 5.438. Its crystals are square tetragonal prisms and pyramids like those of zircon. It is infusible before the blow-pipe, and is insoluble in the simple acids. It dissolves readily in a mixture of nitric and sulphuric acid, and in aqua regia. In nine-tenths of a grain of material, the following constituents were found:

SiO_2	UO_2	ZrO_2(?)	ThO_2	La_2O_3.	Y_2O_3	PbO	FeO	CaO	MgO	K_2O	$(NaLi)_2O$	P_2O_5	H_2O
13.90	22.40	.88	45.30		1.86	3.74	1.15	.59	.10	.43	.68	.67	4.81

The new mineral thus differs from thorogummite in the possession of one molecule of thoria.

Canfieldite is a new germanium mineral from somewhere in Bolivia.[10] Its crystallization is regular, small crystals being bounded by the octahedron and the dodecahedron. The hardness is 2.5, density 6.266, lustre metallic and color black with a purplish tinge. Its streak is grayish black and degree of fusibility 1.5 to 2. Upon analysis, the following result was obtained:

S	Ge	Ag	Fe. Zn.	Ins.		Total
17.04	6.55	76.05	.13	29	=	100.06

which corresponds to the formula Ag_8 Ge S_6. A re-analysis of the Freiberg argyrodite yields results that accord better with the formula above-given than with the formula Ag_4 Ge S_6 proposed for it by its discover, Winkler.[11] Both minerals have the same composition, consequently, since argyrodite is monoclinic, they are dimorphs.

Marshite.—This copper iodide[12] occurs at Broken Hill, New South Wales, as tiny crystals implanted on a siliceous cerussite. The crystals are probably hemihedral-tetragonal. In color they are reddishbrown, in lustre, resinous. They possess an orange yellow streak, are transparent and brittle.

Kehoeite, from Galena, Lawrence Co., S. D., forms seams and bunches in the galena of the Merritt mine. The material is white, amorphous and insoluble in water. Its analysis yielded Headden[13] the following figures:

P_2O_5	SO_3	ZnO	CaO	Al_2O_3	Fe_2O_3	MgO	Cl	H_2O	Ins.	Total
26.76	.50	11.64	2.70	24.84	.78	.08	tr.	31.06	1.76	= 100.02

[10] S. L. Penfield, Amer. Jour. Sci., XLVI, 1893, p. 101.
[11] Jour. f. prakt. Chem., XXXIV, 1886, p. 177.
[12] C. W. Marsh, Proc. Roy. Sci. N. S. W., XXVI, p. 826.
[13] Amer. Jour. Sci., XLVI, p. 22.

corresponding to $R_3(PO_4)_2 + 2 Al_2(PO_4)_2 + 2 Al_2(OH)_6 + 21 H_2O$.

Neptuneite and *Epididymite* are associated[14] with aegirite, arfve-dsonite, eudialyte, etc., near Julianehaab, ˙Kangerdluarsuk, Green-land. The former is found as short, prismatic monoclinic crystals, with a perfect cleavage parallel to ∞P. Their color is black in the larger crystals, but deep red brown in the small ones. Their hardness is 5–6, density, 3.234, and composition:

SiO$_2$	TiO$_2$	FeO	MnO	MgO	K$_2$O	Na$_2$O	Total
51.53	18.13	10.91	4.97	.49	4.88	9.26	= 100.69

These figures correspond to the formula ($\frac{2}{3}$ Na$_2$+K$_2$) Si$_4$O$_9$+($\frac{2}{3}$Fe+ $\frac{1}{3}$ Mn) TiO$_3$. Epididymite is regarded as a dimorph of eudidymite. It occurs in orthorhombic prisms elongated in the direction of their macroaxes. Their analysis: SiO$_2$ = 73.74; BeO = 10.56 Na$_2$O = 12.88; H$_2$O = 3.73, corresponds to the formula for eudidymite, viz.: HNa Be Si$_3$O$_8$. Density = 2.548.

Franckeite, from near Chocaya, in the Animas District, Bolivia, is an associate of the silver ores of the region. It occurs,[15] as a radial, aggre-gate, or as a structureless layer of a dark gray or black substance, that is opaque and soft. Its hardness is about 2.75, and density 5.55. Its quantitative analysis yielded:

Pb	Sn	Sb	S	Fe	Zn	Gangue	Total
50.57	12.34	10.51	21.04	2.48	1.22	.71	= 98.87

while qualitative tests showed it to contain also about .1% of german-iun and a fractional percentage of silver. The mineral is a sulfo-salt of the the the formala: Pb$_2$ Sn$_2$ S$_5$ + Pb$_2$ Sb$_2$ S$_5$. It resembles in appear-ance and in the nature of its components the plumbo-stannite from Moho in Peru, but differs from it in the proportion of its constituents.

Cylindrite owes its name to the cylindrical form that it so commonly assumes. It is described by Frenzel[16] as possessing a dark, lead-gray color and a metallic lustre. It is malleable, has a hardness of 2.5–3, and a density of 5.42. It occurs in cylindrical bodies imbedded irregu-larly in a granular lamellae mass of the same substance. An analysis of the mineral gave:

Pb	Ag	Fe	Sb	Sn	S	Total
35.41	.62	3.00	8.73	26.37	24.50	= 98.63

[14] G. Flink, Geol. För. Förh., XV, 1893, p. 195.

[15] A. W. Stelzner, Neues Jahrb. f. Min., etc., 1893, II, p. 114.

[16] Ib., 1893, II, p. 125.

corresponding to $Pb_6 Sb_2 Sn_6 S_2$. The mineral is easily decomposed by hot hydrochloric and nitric acid, but is scarcely affected by cold hpdrochloric acid. Like franckeite and plumbostannite, it is a South American mineral, occurring, as it does, at the Mina Santa Cruz, Poopó, Bolivia.

Hantefeuillite accompanies crystals of apatite, pyrite, iron and monazite at the apatite mine at Odegärden, Bamle, Norway. It is found in the greenish nodules composed of wagnerite and apatite, that are scattered through the apatite veins cutting gabbro. Michel[17] describes it as forming transparent, colorless monoclinic crystals radically grouped. Its hardness is 2.5 nnd density 2.435. The crystals are all tabular in habit, being elongated parallel to *c*, and flattened to $\infty P \bar{\infty}$. Their optical axes lie in the latter plane, and their optical angle has a value—2Vna = 54° 23'. An analysis gave $P_2O_5 = 34.52$; $MgO = 25.12$; $Cao = 5.71$; $H_2O = 34.27$, corresponding to $(Mg\ Ca)_3\ (PO_4)_2 + 8H_2O$, which is the guano mineral bobierrite in which Mg has been in part, replaced by Ca.

Chondrostibian, as its name indicates, is an antimony mineral occurring in grains. It is reported by Igelström[18] from the famous manganese mine Sjogrufran, Grythyttan, Sweden. It is found disseminated as grains through barite, which, with calcite and tephroite, forms a cryptio-crystalline mass. These grains constitute nearly 50% of some of the barite plates to which they impart a brownish tinge. The mineral itself is yellowish-red in color, though in large pieces it appears dark brownish red. It is weakly magnetic, and yields, upon analysis, figures indicating the following composition:

Sb_2O_5	As_2O_5	Mn_2O_3	Fe_2O_3	H_2O	Total
30.66	2.10	33.13	15.10	19.01 =	100.00

corresponding to $3R_2O_3\ Sb_2O_5 + 10\ H_2O$.

[17] Bull. d. l. Soc. Franç d. Min., xvi, p. 88.
[18] Zeits. f. Kryst., xxii, p. 43.

Reprinted from The American Naturalist, January 1st, 1894.

MINERALOGY AND PETROGRAPHY.[1]

Globular Granite in Finland.—An occurrence of spherical granite is reported by Frosterus[2] from the southern and eastern portions of Borga in South Finland. In the midst of a number of knolls of red or gray microcline granite, is one in which spherical nodules are plentiful. Of the rock forming this knoll there are two varieties distinguished by the difference in size of their nodules. In one the nodules are small and consist of a light covered zone surrounded by a dark periphery composed of two or three concentric biotite shells. The kernel is a granular aggregate of oligoclase, some microcline, a little quartz, and considerable biotite toward the center. The rock enclosing the nodules is a dark gray granite in which quartz and microcline are more abundant than in the nodules. In the second variety of rock the nodules are large. Their kernels are like the small nodules described. Around these is usually a narrow band of feldspar and around this a zone of mica. The rock in which the spherules lie is a grayish red granitite.

After investigating carefully the relations of the minerals in the nodules to each other and the relations existing between the nodules themselves, the author concludes that the spherules existed as plastic bodies in the rock magma while this was still liquid. When in contact with each other the nodules are often distorted, whereas at other times they are broken across. It is believed that the mica and other more basic components first separated in the form of a shell enclosing some of the rock's magma, that afterward gave rise to the granular nucleus upon cooling. The nodules are thus looked upon as basic concretions, and since they are distributed through a few restricted areas only, they are thought to form basic " Schlieren." The author's article is well illustrated by several handsome plates.

The Inclusions in the Basalts of the Oberlausitz.—A further study of the granite inclusions in the basalts of Oberlausitz by Beck[3] adds a few items of information concerning the contact action between volcanic rocks and their included fragments. On the Hirschberg the granite inclusions in nepheline-basalt have had pro-

[1] Edited by Dr. W. S. Bayley, Colby University, Waterville, Me.
[2] Minn. u. Petrog. Mitth., XIII, p. 177.
[3] Minn. u. Petrog. Mitth., XIII, p. 231.

duced in them spinels and augite.   The dyke melilite-nepheline-basalt
near Kemnitz becomes porphyritic around the inclusions.   Spinels and
augite are again the principal new products formed in the granite, but
in addition to these glass nodules containing chalcedony and tridymite are also found in the inclusions.   On the basalt side of the contact nepheline is lacking and feldspar takes its place, while the olivine
of the original rock is broken and corroded.   Around a few of the
inclusions a mineral of the hauyne group has developed.   The nepheline-basalt of the Spitzberg near Paulsdorf, contains a very large number of included fragments, around which the course of the contact
processes may be easily studied.   Around some of them is an isotropic
glass containing microlites and trichites, while one large inclusion
made up of many fragments is discovered under the microscope to
· have its pieces cemented by glass in which are feldspar and quartz
fragments, and now and then small crystals of augite forming 'crowns'
around the quartzes, besides biotite, granular colorless olivine and crystals of cordierite, which are always associated with magnetite.   As the
distance from the inclusion increases, the quartz and feldspar gradually
disappear, augite increases in quantity and olivine of the basalt type
becomes prominent.   The rock then differs from the normal nepheline-
basalt mainly in containing feldspar and in the absence of nepheline.
Of course, at a greater distance from the inclusion, the rock assumes
its normal composition.

Thermometamorphism around the Shap Granite.—In a
paper published some two years ago and abstracted in the Bulletin of
the Geological Society of America[4], Messrs. Harkes and Marr[5] discussed the interesting effects produced upon andesite and rhyolitic
lavas and tufas and upon limestones and slates by the intrusion through
them of a great mass of granite at Shap Fell, in the Lake District, England.   The same gentlemen return[6] to their study in a late paper, sup-
plementing and correcting their former statements.   They find in addition to the andesites and rhyolites, sheets of basalt or of a very basic
andesite, containing monoclinic and orthorhombic pyroxenes, and like
the other lavas characterized by an abundance of vesicles filled with
products of weathering.   These have suffered contact alteration to a
greater extent than have the primary constituents, though all have

[4] Bull. Geol. Soc. Amer., Vol. III, 1892, p. 16, cf. AMERICAN NATURAILST,
1892 p. 847.
[5] Quart. Jour. Geol. Soc., XLVII, 1891, p. 266.
[6] Quart. Jour. Geol. Soc,, 1893, XLIX, p. 359.

been affected near the contact with the granite. Green hornblende, brown mica, colorless pyroxene, epidote and sphene are the most conspicuous new minerals formed. These lie in a clear, granular mosaic, which may consist of newly developed quartz and feldspar. The components of the vesicles have in most cases given rise to a mixture of hornblende and quartz, but in other cases a little calcite may remain unaltered in the center of larger vesicles, while surrounding it are usually hornblende, colorless pyroxene, quartz and epidote, and sometimes in addition, zonal garnets, sphene and a few other minerals. The feldspar found within the vesicles of metamorphiosed andesites is thought by the authors to be the result of the weathering of these rocks rather than a product of contact action. In concluding their paper some interesting thoughts are suggested as to the source of the materials producing contact minerals. It is known that limestones when pure may recrystallize as marbles without the production of contact minerals, but that when impure the silica in the impurities may (and generally does) release the carbonic acid and recrystallize with the calcium as silicates. In some of the vesicles of the rocks around the Shap granite, however, the calcite has recrystallized, with the formation of silicates only around the edges of its mass, proving plainly that silica was obtained for the production of the silicates only by the calcite immediately in contact with the silicates. The conclusion is that in cases of thermometamorphism no transference of material takes place within the mass of the altered rocks except between closely adjacent points. In the production of the lime silicates studied, the interchange of lime and silica is estimated to be limited to a distance of $\frac{1}{10}$ of an inch. Other observations indicate the correctness of this conclusion.[1]

Petrographical News.—In the Obersweiler gneiss of north Vogesen are dykes of basic rocks that Andreae and Tenne[2] identify as hornblende kersantites. They consist of a panidiomorphic aggregate of plagioclase, green hornblende, a little mica, quartz, apatite, etc. Other dykes of the region are quartz-melaphryes of the navite type. The quartz is undoubtedly original. Its grains are much corroded and the resorption rims around it are composed of augite and glass. The rock is interesting as the first recorded example of a dyke rock corresponding to the volcanic quartz-basalts.

The porphyritic granite of northern Lausitz contains large numbers of apatite crystals, sometimes as many as a hundred in a single

[1] See also Journ. of. Geol., Vol. I, p. 574.
[2] Zeits. d. deutsch. geol. Ges. 1892, p. 824

thin section.   As large as is this number it is exceeded in sections of
the basic concretions of the rock from Niedersteina.   These concretions according to Hermann[9] are made up largely of hornblende and
cordierite, and thousands of apatites, sometimes reaching 1200 in a
single section.   The interesting features of these apatites is not, however, their number, but their forms.   In many cases they are skeleton
crystals whose many branches are parallel like the teeth of a comb.

The Hour-Glass Form of Augite.—This well known form of
augite, according to Blumrich,[10] is usually connected with zonal growth
in the mineral, and is limited in its occurrence to the pyroxene of alkaline rich magmas.   It is found not only in augite, but also in other
minerals forming colored isomorphorus mixtures.   The hour-glass form
owes its existence to the fact that different crystallographic faces in a
growing mineral attract molecules of different chemical compositions,
which by addition to the attracting faces build out these faces with
differently colored substance.   The structure is certainly not due to the
filling in of the outlines of skeleton crystals, as has often been assumed.   Zonal bands extend uninterruptedly through both dark and
light areas in the crystals, hence the materials of both must be of the
same age.   The one cannot have been a later deposition than the
other.   Pelikan[11] in confirmation of Blumrich's view, calls attention to
the fact that if strontium nitrate crystals be allowed to grow in certain colored solutions, they become colored in areas distributed in
accordance with the faces by which the crystals are bounded.   The
central cores of chiastolite crystals, Becke ascribes in a similar manner to the attractive influence of the end faces of the crystals upon the
material added during growth.

The Effect of Impurities in Crystallizing Solutions.—It
has long been known that the habit of crystallization assumed by a
substance depends in large measure on the medium from which it cry-
tallizes.   Araganite, for instance, will separate from certain solutions, while from others calcite is precipitated.   Vater[12] has conducted
a series of experiments with calcium carbonate, allowing this substance
to crystallize from various solutions under different conditions; and
has reached some interesting conclusions.   The ground rhombohedron

[9] Neues Jahrb. F. Min., etc., 1893, II, p. 52.
[10] Minn. u. Petrog. Mitth., XIII, p. 239.
[11] Ib. XIII, p. 258.
[12] Zeits. f. Kryst. XXI, p. 433 and XXII, p. 209.

of calcite separates from all solutions of pure carbonate in dilute carbonic acid at low temperatures. In general, under different conditions of formation, differently habited crystals are produced. Moreover, different proportions of impurity in the solution affect differently the resulting crystals, as well as the rapidity with which they grow. Contrary to the prevalent belief, however, the presence of calcium bicarbonate in a solution of the mono-carbonate exerts but little influence upon the complexity of the calcite crystals formed. The article is long, and is a thorough discussion for the subject treated.

North Carolina Quartz Crystals.—Gill[13] supplements Von-Rath's study of North Carolina quartz crystals by describing some new forms and giving the results of etching spheres made from simple left-handed crystals with hydrofluric acid and hot sodium carbonate. The conclusions of his crystallographic study are to the effect that the mean of the measurements of 38 crystals give an axial ratio $a : c = 1 : 1.1018$. This ratio, which is larger than usual for quartz, is ascribed to the lengthening of the c axis brought about by impurities included within the crystals. All the crystals investigated were smoky quartzes, whose axial ratio approaches that of the Swiss crystals, and is larger than that of the Riesengrunde occurrences $(1 : 1.0996)$. The crystallazition is trapezohedral-tetartohedral, which may be best regarded as *a combination of trapezohodral hemihedrism* and hemimorphism with respect to the lateral axes. The author notes the effect of various influences upon the development of the planes observed on quartz, and closes his paper with a discussion of crystal structure. The properties of quartz are explained upon the assumption of a molecule of SiO_2 in which Si is in the center of a regular tetrahedron, from whose upper and lower edges the oxygen exercises its influence.

Two New Books.—Hatch's mineralogy[14] is an elementary text book for the use of beginners in the study of minerals. The book begins with a very elementary treatment of the systems of mineralogy based in the notion of symmetry. It defines the terms made use of in describing the physical properties of minerals and ends with seventy-five pages on systematic mineralogy. The classification used is an arbitrary one—the rock-forming minerals being first discussed, then the ores, next the salts and other useful compounds and finally the

[13]Zeits. f. Kryst., XXII, p. 97.
[14]Mineralogy by F. H. Hatch, London, Whittaker & Co., 1892. Pps. viii and 224. Ills.

gems. The descriptions are clear but very brief and the illustrations in the text are well selected. The little volume is one of the best of its kind, though this is but scant praise.

Gregory's translation[15] of Loewinson-Lessing's Tables for the Determination of Rock-Forming Minerals, adds another to the number of books that are supposed to aid the student in the rapid determination of the most common constituents of rocks. The tables are intended to lead their user to the *name* of the mineral whose characteristics he has observed under his microscope. It is a " guide to the identification of minerals, rather than a summary of their properties. " The plan made use of in the construction of the tables reminds one of the schemes familiar to the determinative botanist. Habit, color, lustre, character of double refraction, etc., serve to place the minerals in different groups, from which one whose name is sought is selected by its special characteristics. The tables appear to fill a want, but only constant use in the laboratory will prove whether or not they will assist the student to the extent hoped by the author.

Mineralogical News.—*Azurite* with the habit of Chessy crystals and large *cerussites* prismatic in the direction of the brachydiagonal are mentioned by Molengraff[16] from Willow's silver mine near Pretoria in the Transvaal. On the former the three new planes $\frac{1}{10}$ P$\stackrel{>}{\infty}$,—2 P$\stackrel{>}{?}$ and $\frac{1}{4}$ P$\stackrel{>}{?}$ occur.

On three highly modified crystals of *phosgenite* from Monte Poni, Sardinia Goldschmidt[17] has discovered the new forms P$\frac{1}{4}$ and 3 P$\frac{1}{4}$ The distribution of the more common faces seems to point to a trapezohedral symmetry for the crystals, but no circularly polarizing effects could be detected in them. The axial ratio determined from the mean of the best measurements is $a : c = 1 : 1.0888$.

An analysis of *jarosite* from the cavities of the auriferous quartzite of the Buxton Mine, Lawrence Co., S. D., has been made by Headdon.[18] His results are:

SO_3	As_2O_5	Fe_2O_3	CaO	Na_2O	K_2O	H_2O	Total
30.29	2.51	49.28	.42	4.62	1.57	11.24	= 99.93

[15]F. Loewinson-Lessing's: Tables for the Determination of Rock-Forming Minerals. Translated by J. W. Gregory, With a chapter in the petrological microscope. London & N. Y., MacMillan & Co., 1893. Pp. 55.

[16]Zeits. f. Kryst., XXII, p. 156.

[17] Ib., XXI, p. 821.

[18]Amer. Jour. Sci., XLVI, 1893, p. 24.

Three fragments of *powellite* have been obtained by Koenig and Hubbard[19] from the south Hecla copper mine in Houghton Co., Mich. The mineral has a density of 4.349. Its composition was found to be:

MoO$_3$	WO$_3$	CaO	MgO	Fe$_2$O$_3$	SiO$_2$	Cu	Total
67.84	1.65	27.30	.16	.96	1.52	tr	= 99.43

Native *lead* is reported by Kempton[20] as occurring in thin scales and pellets, some of which approach rectangular forms, in a gangue of pyroxene of a pale green color. It is associated with iron oxides and calcite. The location given is near Saric, Sonora, Mexico.

Methods and Instruments.—Federow[21] in a recent article elaborates a new universal method for the measurement of crystals, suggests a new system for crystallographic nomenclature and illustrates a new method of projecting crystal planes, and determining by graphical means their symbols. The universal goniometer used in his investigations is described at length and pictured in detail. The author illustrates also the application of his method to studies in optical crystallography. He describes two models of universal microscope stages, constructed for the purpose of enabling the observer to revolve the object under investigation in two directions. The plagioclases are studied and it is shown that the labor of determining their nature is much reduced by the method of work suggested by the author. The paper is an important one and and one well worthy of close study.

Czapski[22] suggests the use of the iris diaphragm between a condenser of moderate strength and the stage of the microscope for the rapid interchange of parallel and converged light, and also the use of the same appliance in the ocular tube of the instrument for the isolation of the axial figures of very small crystals.

G. Friedel[23] gives a new method for determining the value of the double refraction in thin sections of minerals that seems to be simple in its application.

Goldschmidt[24] and Jolles[25] discuss two proposed methods for projection of crystal forms. Jolles article is illustrated by five plates and sixty figures.

[19]Ib., XLVI, 1893, p. 356.
[20]Science, June 23, 1893, p. 345.
[21]Zeits. f. Kryst., XXI, p. 574 and XXII, p. 229.
[22]Ib., XXII, p. 158.
[23]Bull. Soc. Min. Franc., XVI, p. 19.
[24]Ib., XXII, p. 20.
[25]Ib., XXII, p. 1.

MINERALOGY AND PETROGRAPHY.[1]

The Eruptive Rocks of Cape Bonita, Cal.—The eruptive rocks forming the main mass of Cape Bonita, the northern Cape separating San Francisco from the Pacific Ocean, are spherical basalts and diabases, in addition to basic tuffs. The basalt is remarkable for the great spheroidal masses that characterise it. In many places the entire rock-mass is a closely packed aggregate of large bolster-like bodies, whose cross-section is approximately circular. These consist of a compact amygdaloidal rock, made up of lath-shaped plagioclases lying in a glassy base. In all cases the rock of the spheroids is much altered, and is of the same composition in the interiors as on the peripheries of the bodies. In a few cases augite may be detected as small grains that are younger than the plagioclases, but the rock on the whole is very uniform in character. The diabase is more interesting petrographically. It is younger than the basalt and has intruded this rock. Besides the usual constituents of diabase it contains iddingsite in large, rounded, idiomorphic forms. The augite varies in color from nearly colorless to a deep violet red, the latter varieties possessing a pleochroism in yellowish green and violet red tints. A qualitative test showed the presence of titanium. Sometimes the augites of different colors are intergrown, when they are optically continuous, and not infrequently the mineral is intergrown with brown hornblende. The outlines of the iddingsite are strongly suggestive of olivine. It was one of the earliest separations from the magma, being included in the augite and in the hornblende. Its own enclosures are magnetite and chromite or picotite. In some phases of the rock both green and brown hornblende are present. Both of these are regarded as original and as of the same age as the augite, for they are frequently intergrown with the pyroxene as well as with each other. In one place the diabase is variolitic, with variolites composed of tiny brushes and crystallites of various minerals, lying in a microlitic diabasic groundmass. Iddingsite occurs both in the groundmass and in the varioles. The pyroclastic rock associated with the basalt and the diabase is probably an ash of a basaltic character. Some of its component fragments resemble closely the material of the spheroidal rock. Analyses of the rocks discussed are given by Mr. Ransome,[2] in a recent number of the University of California Bulletin.

[1] Edited by Dr. W. S. Bayley, Colby University, Waterville, Me.
[2] Bull. Geol. Dept. Univ. Cal., Vol. 1, p. 71.

Lamprophyres near the Shap Granite Mass.— Near the Shap granite in the North of England there are numerous dykes of minette and kersantite that are believed by Harker[3] to be the dyke facies of the granite, just as fourchite and ouachitite are regarded by Rosenbusch as dyke facies of eleolite-syenite. These lamprophyres contain many rounded blebs of quartz and corroded crystals of orthoclase, both of which appear to owe their present shapes to resorption processes, since both minerals are surrounded by resorption borders. The dyke rocks are thought to be genetically connected with the granite because of their age and distribution, and because of the fact that they contain the quartz and orthoclase above referred to, and also sphene, which is a characteristic component of the granite. A study of the literature of the lamprophyres shows that these rocks are often associated with granites, and hence Harker believes that the group may be discovered to be genetically related to this group of plutonic rocks. A special feature of the lamprophyres pointed out by the author is that while the total alkalies in them is about equal in amount to the sum of the alkalies in the associated granite, the potash in the former always bears a larger ratio to the soda than it does in the latter rock. It is suggested that the granite and the lamprophyres are portions of the same magma that became differentiated by gravity. From the supernatant layer, which was acid, quartz and orthoclase separated and then settled down into the lower basic portions of the mass. These were then partially dissolved, the solution of the orthoclase accounting for the large proportion of potash in the lamprophyres. In a later paper the author[4] argues against the view of Diller and Iddings that the sporadic quartzes in certain basalts and other basic rocks are the result of crystallization under other than the normal conditions. He thinks that in all these cases the quartz may have originated as outlined above.

The Geology of Conanicut Island, R. I.—The carboniferous phyllites of Conanicut Island in Narragansett Bay are cut by a mass of coarse-grained muscovite granite porphyry that has produced contact effects in the surrounding sedmentaries.[5] The granite, which exhibits many evidences of its intrusive nature, was regarded by Dale[6] as a metamorposed clastic rock, forming the lowest member of the bedded series at this place. The phyllites near the contact with the granites

[3] Geol. Magazine, 1892, IX, p. 199.
[4] Ib. IX, p. 485.
[5] L. V. Pirsson. Amer. Jour. Sci., 1893, XLVI, p. 363.
[6] Proc. Bos. Soc. Nat. Hist., 1883, XXII, p. 179.

have been changed into hornstones and knotty schists. Besides the granites the only other intrusives cutting the slates are two dykes of minette, both of which show the effects of pressure. One of the dykes consists essentially of orthoclase and two generations of biotite. It contains also apatite and zircon and large quantities of plagioclase and calcite. In the squeezed phase of the rock the biotite has been changed to chlorite. The material of the second dyke differs from that of the first one, only in that it has been more thoroughly squeezed and consequently has suffered greater alteration.

Petrographical News.—Sears[7] finds that the porphyritic feldspar in the rock from Marblehead Neck, Mass., called by Wadsworth[8] trachyte, are anorthoclases, and that much of the feldspar of its groundmass is of the same nature, consequently the rock is a keratophyre. Analyses of the rock and of one of its phenocrysts follow:

	SiO_2	TiO_2	Al_2O_3	Fe_2O_3	FeO	MnO	CaO	MgO	K_2O	Na_2O	P_2O_5	H_2O
Rock	70.23	.03(?)	15.00	1.99		.24	.33	.38	4.99	4.98	.06	2.19
Felds.	65.66		20.05	tr.	tr.	.13	.67	.18	6.98	6.56		.41

The report of the State Geological Board of Michigan[9] contains brief microscopic descriptions of certain eruptive, sedimentary and schistose rocks of the Upper Peninsula by Drs. Patton and Lane. Among the former are described granites, syenites, serpentine and lamprophyres. Among the sedimentaries graywackes, quartzites and slates, and, among the foliated rocks, amphibolites and hornblende schists. The amphibolites are principally altered diabases. Quartz diabases are mentioned by Lane as existing in dykes cutting graywackes and slates that are sometimes changed on the contact into spilosites, and quartzites that are altered near the intrusive into Lydian stone. Dr. Wadsworth, in the same volume, gives an outline scheme of his classification of rocks (eruptive and sedimentary), the principles of which were first ennunciated at length in his Lithological Studies.[10]

Graeff[11] has found, in an old hand specimen of tephrite from Horberig in the Kaiserstuhl, a holocrystalline basic concretion with a structure approaching that of theralite.

[7] Bull. Mus. Comp. Zool., Vol. XVI, p. 167.
[8] Proc. Bost. Soc. Nat. Hist., XXI, p. 288.
[9] Rep. State Board of Geol. Survey for 1891-92. Lansing, 1893.
[10] Mem. Mus. Comp. Zool., 1884, XI.
[11] Versamm. Oberrh. Geol. Ver. Ber., XXVI, 1893.

A modification of the microchemical method for determining iron in minerals is given by Lemberg.[12] It consists in producing Turnbull's blue from the ferrous sulphide precipitated on the mineral in question.

Alurgite and Violan from St. Marçel.—Among the minerals from the Manganese mines of St. Marcel, Piedmont, *alurgite* and *violan* haae always excited considerable interest because of their rich color and their variety. The alurgite was described by Breithaupt as a deep red mica. Penfield[13] has recently obtained a sufficient quantity of the material for study. He describes it as monoclinic in crystallization and micaceous in habit. Its cleavage plates are flexible and somewhat elastic. It is biaxial with $2 E_{na} = 56° 32'$ (average) and its dispersion is $\mathfrak{r} > \mathfrak{v}$, but often plates show a uniaxial optical figure, due, as the author supposes, to twinning. The mica is one of the first order, and in spite of its dark color, its pleochroism is very slight. Density $= 2.835$ —2.849. $H = 3$. Composition:

SiO$_2$	Al$_2$O$_3$	Fe$_2$O$_3$	Mn$_2$O$_3$	MnO	MgO	K$_2$O	Na$_2$O	H$_2$O	Total
53.22	21.19	1.22	.87	.18	6.02	11.20	.34	5.75	= 99.99

In the formula $H R_2 (Al OH) Al Si_4 O_{12}$, $R = K$ and $Mg OH$. Alurgite is thus a distinct species, which is more nearly allied to lepidolite than to muscovite, although it is a potash mica. The alurgite is associated with a jadeite composed largely of a soda-rich pyroxene that is pleochroic in pale rose and pale blue tints. Its density is 3.257—3.382, and composition (mean of two analyses):

SiO$_2$	Al$_2$O$_3$	Fe$_2$O$_3$	Mn$_2$O$_3$	MnO	MgO	Ca$_2$O	Na$_2$O	K$_2$O	Ign	Total
54.59	9.74	11.99	1.06	.58	5.03	7.24	9.32	.24	.37	=100.16

corresponding to $Na R (SiO_3)_2$, in which $R = Al, Fe''', Mn'''$. The mineral occupies about the same position in the pyroxene group as glaucophane does among the amphiboles. In composition it agrees most closely with the chloromelanite from Mexico analysed by Damour.[14]

For purposes of comparison with this pyroxene, the author analysed a specimen of violan whose density was 3.272 — 3.237, with this result.

SiO$_2$	Al$_2$O$_3$	Fe$_2$O$_3$	Mn$_2$O$_3$	MnO	MgO	CaO	Mn$_2$O	K$_2$O	Ign	Total
53.94	1.00	.86	.88	.36	16.63	23.80	1.22	.05	.66	= 99.44

The figures indicate a mixture of the diopside, jadeite and acmite mole-

---

[12] Zeits d. deuts. geol. Ges., 1892, p. 823.

[13] S. L. Penfield. Amer. Jour. Sci. XLVI, p 288.

[14] Bull. Soc. Min. d. Franc, IV, 1881, p. 157. Cf. also foot-note No. 30.

cules in the proportions 90.8 : 4.1 : 2.4, with the addition of 2.7% of the molecule Na Mn $(SiO_3)_2$. The mineral is essentially a blue variety of diopside, differing from the anthochroite of Igelström[15] and from the blue pyroxene of Merrill and Packard.[16]

Zonal Plagioclase.—Herz[17] has shown by a study of the position of axial planes in successive zones of zonal plagioclase, and by the values of the respective cleavage angles, that the zonal banding in this mineral is due to the concentric growth of envelopes of different composition. The axial planes and the cleavage angles always correspond with the extinction angles in the corresponding band. It had been suggested by Grosser that the regular decrease in the extinction of the shells of a zonal plagioclase is due to difference in the orientation of the successive envelopes and not to a difference in their chemical composition. Herz's work proves conclusively that the decrease in the value of the extinction is not due to differences in orientation of the same chemical substance.

Hercynite in Gabbro.—Small octahedra and large irregular masses of the green spinel hercynite occur in an altered gabbro at Le-Prese, in the Valtellina, Switzerland. According to Linck,[18] it is found as irregular granular masses within the rock, and as small octahedral crystals enclosed in its plagioclase and associated with corundum sillimanite and biotite. The spinel includes small quantities of biotite, small plates of ilmenite, resembling the plates in hypersthene, a little pyrite, etc. An analysis of tolerably pure material yielded :

SiO_2	Al_2O_3	Fe_2O_3	MgO	FeO	Total
1.59	59.62	3.10	9.38	25.30 =	98.99

which corresponds to (Fe Mg) Al_2O_4 in which Fe : Mg $= 3:2$.

Optical Constants of Topaz.—Four Japanese topaz crystals and one crystal of the same mineral from New South Wales are described by Hahn,[19] and some of the optical constants of the former have been determined. One of the crystals from Otamjama near

[15] AMERICAN NATURALIST, 1890, p. 74.
[16] Ib., 1892. p. 848.
[17] Min. u. Petrog. Mitth. XIII, p. 341.
[18] Sitzb. d. Kön-preuss. Akad. d. Wiss. zu Berlin. Phys.-Math.-Classe., 1893, p. 47.
[19] Zeits. f. Kryst., XXI, p. 334.

Kioto, has the following refractive indices and optical angles for yellow light: $\beta = 1.6182$, $\gamma = 1.6252$, 2 V = 62° 40′, 2 E = 114° 31′. The crystal from New South Wales has 2 E = 113° 18′.

Mineralogical News.—Stöver announces the discovery of fine *celestites* in the Jurassic schists of Brousseval in France. Their axial ratio is .7803 : 1 : 1.2826, and index of refraction for sodium light = 1.6235. The crystals are one centimeter in length, and are elongated parallel to ă. Similarly habited crystals occur also in the marl of Ville-sur-Sault. The axial ratio of these is .7806 : 1 : 1.2797, and density = 3.991.

Rheineck[20] has made another attempt to calculate from the published analyses general formulas for *tourmaline* that will not only represent the composition of all varieties of the mineral, but which will also express its relationship with micas. He concludes that there are two alkaline varieties, viz.: $Al_4 Si_3 B H_3 O_{15}$ and $Al_4 Si_3 B_2 H_4 O_{17}$, and two magnesium varieties, $Al_4 Si_5 B_2 Mg_4 O_{25}$ and $Al_4 Si_5 B_3 Mg_3 O_{23}$, by whose intermingling all other varieties are formed.

Several crystallographic observations of Baumhauer[21] are of interest. A yellow *diopside* from the Canton of Graubünden (Grisons), Switzerland, has an axial ratio $a : b : c = 1.0918 : 1 : .5879$, with $\beta = 74°$ 12′ 15″. *Binnite* crystals from Infeld in the Binnenthal are certainly tetartohedrally hemihedral, as the author has succeeded in finding upon them, well-developed, the planes $\frac{0}{7}$ and $\frac{20 2}{7}$.

Oebbecke[22] mentions the occurrence of *topaz* with feldspar, apatite, tourmaline, fluorite, etc, at Epprechtstein and its existence in the granite of the Gregnitzgrund in the Fichtelgebirge.

The *arsenopyrite* of Weiler in Alsace occurs in an arkose from which Scherer[23] has obtained crystals sufficiently large for measurement and analysis. These crystals are prismatic in habit, and have an axial ratio $a : b : c = .6734 : 1 : 1.1847$. A mean of two analyses gave figures corresponding to Fe : S : As = 1 : .9933 : .9751.

Mallard[24] has come into the possession of some beautiful little crystals of *periclase* that were found implanted on a white compact crust produced in the culcination of some of the Stassfurt materials.

Several twins of *aragonite* from the tunnel of Neussargues in Cantal,

[20] Zeits. f. Kryst., XXII, p. 52.

[21] Ib., XXI, p. 200.

[22] Ib., XXII, p. 278.

[23] Ib., XXII, p. 62.

[24] Bull. Soc. Franc. Min., XVI, p. 18.

France, are reported by Gonnard[25], and some fine crystals of *pinite*[26] from Issertaux, near St. Pardoux in the Auvergne.

Miscellaneous.—In his development of the theory of the constitution of the *micas*, Clarke[27] has reached the problem of the lithium members of the group. This he solves by supposing lepidolite to be an admixture of the simple molecules Al F, Si, O, R,', in which R' is principally lithium, and Al, (SiO,), R,', in which R,' may be either K,H or KH,.

Retgers[28] suggests molten phosphorus and a solution of phosphorus in CS, as media for use in determining the indices of refraction in highly refracting substances. A tiny fragment of the phosphorus may be melted between two object-glasses, when it spreads as a thin sheet between them, and, upon cooling, remains transparent. Its refractive index is 2.144. That of a saturated solution of the substance in CS, is 1.95.

Some time ago, Damour[29] suggested the name *chloromelanite* for one of the varieties of jade found in ancient implements. He discovers now that the material contains garnets and pyroxene. It thus resembles the rock eclogite. The pyroxene from a Mexican specimen is composed as follows:

SiO,	Al,O,	Fe,O,	CaO	MgO	Na,O	Total	Sp. Gr.
56.57	17.21	8.86	4.44	2.12	10.70 =	99.90	3.37

Nordenskjöld[30] has begun the study of snow crystals. The first contribution to his discussion is a series of handsome photographs of a large variety of flakes, including prismatic, stellar and other forms some of which contained liquid enclosures at the time of their fall.

[25] Ib., XVI, p. 10.
[26] Ib , XVI, p. 16.
[27] Bull. Am. Chem. Soc., XV, May, 1893.
[28] Neues Jahrb. f. Min., etc., 1893, II, p. 130.
[29] Bull. Soc. Franc. Min., XVI, p. 57. Cf. also foot-note No. 14.
[30] Ib., XVI, p. 59.

MINERALOGY AND PETROGRAPY.[1]

Eleolite Rocks from Trans-Pecos Texas.—In a recent report on Trans-Pecos Texas Osann[2] gives a few brief notes on the igneous rocks of the region. The most interesting points in the article, which, on account of the short time allowed the author to prepare it, is little more than a collection of notes, refer to the alteration of limestones by granite and the production of a rock composed almost exclusively of calcium silicates; to the existence of eleolite syenites and phonolites in the Davis Mountains; to the occurrence of a tourmaline schist in the Van Horn Mountains, and of altered diabases and squeezed porphyries in the Carriso Mountains. The eleolite syenite is a fine grained, light colored rock with the typical trachytic structure. It contains orthoclase, eleolite and olivine as phenocrysts and sodalite, aegyrite, malacolite, hornblende. arfvedsonite and the rare minerals ainigmatite, laavenite and pyrrhite in its groundmass. The olivine is nearly colorless in thin section. It usually plays the part of a nucleus around which the other dark components have crystallized. The pyroxene occurs in two generations. The amphiboles are also in two generations, and often these and the pyroxenes are intergrown with their *c* axes and clinopinacoids coinciding. Ainigmatite is common in the rock, laavenite and pyrrhite are rare. The phonolites fall into two types. Those of the first type are characterized by their fine grain, by the abundance of needles and grains of aegyrite in their groundmass, and the absence from them of amphibole and other accessory components. In the rocks of the second type are a few phenocrysts of feldspar and of nepheline, the latter of which are often bordered by a dark corona of bisilicates. The most prominent of these are aegyrite and malacolite among the pyroxenes and among the amphiboles a variety with a strong pleochroism as follows: *A*=dark greenish blue; *B*=dark grayish brown; *C*=light yellowish brown. Cutting the eleolite syenite are dykes of tinguaite, monchiquite, alnoite, ouachitite, and a rock to which the author gives the name paisanite, since it was found in Paisano Pass in the Davis Mountains. This new rock consists of a few phenocrysts of quartz and of sanidine in a dense white matrix spotted with blue hornblende whose optical properties show it to be riebeckite. The white matrix is composed of

[1] Edited by Dr. W. S. Bayley, Colby University, Waterville. Me.
[2] Fourth Ann. Rep. Geol. Survey of Texas, p. 123.

intergrowths of albite and orthoclase cemented by granophyric quartz. It is unfortunate that the author cannot further pursue the studies so auspiciously begun.

The Differentiation of Rock Magmas.—In a recent number of the *Journal of Geology* are two contributions relating to the theory of the differentiation of rock magmas. One, by Iddings,[3] is a simple statement of the nature of the phenomena that have led to the proposal of the theory. The article does not discuss the causes of the differentiation of magmas except in general terms, but it deals with the facts that seem to indicate that such a differentiation of a homogeneous magma into unlike parts is alone capable of accounting for the great differences observed in the various rocks emanating from a single volcanic center, and in different portions of the same rock mass. The second article, by Backtsröm[4], was written to call attention to the difficuly of explaining magmatic differentiation upon Soret's principle, which applies, so far as we know, only to dilute solutions, and effects only the proportions existing between the solvent and the dissolved body in different portions of a solution. The author prefers to consider rock magmas as mixtures of liquids, some of which are less soluble in others at certain temperatures than at certain different temperatures. Hence if a homogeneous magma cools to a temperature when some of its constituents become difficultly soluble in the mixture of the others, it will become separated into parts possessing different compositions—liquation will ensue. Thus basic concretions are sometimes formed in acid rocks, and the acid and the basic lavas of Iceland occur in numerous flows, side by side, while intermediate rocks are absent.

The Old Volcanics of South Mountain, Pennsylvania.— Miss Bascom[5] has examined with great thoroughness the acid volcanics of South Mountain, Pa., whose existence was made known to the geological public a year[4] ago, and has described briefly the results of her study. These volcanics exhibit many of the features of modern rhyolites in spite of the fact that they have undergone profound alteration since their eruption. Fluidal, micropoicilitic, spherulitic, axiolitic and lithophysal structures are noticed in the various speci-

[3]Jour. Geol., Vol. I, p. 833.
[4]Ib., Vol. I, p. 773.
[5]Jour. Geol., Vol. I, p. 813.
[4]Amer. Jour. Sci., XLIV, p. 482.

mens; perlitic parting is occasionally detected in them; amygdaloidal phases are not uncommon, while taxitic and trichytic structures are frequently met with. The original components of many of the South Mountain rocks have entirely disappeared and in their place are now found only quartz, epidote, magnetite and leucoxene. These minerals are evidently secondary and yet in some specimens they are associated in micropoicilitic intergrowths, thus indicating to the author the secondary origin of this structure in the present instance. The spherelites in the rocks under consideration are often imbedded in a base that was formerly a glass, though it is now a holocrystalline quartz–feldspar mosaic, which must necessarily be of the nature of a devitrification substance, since the mosaic is crossed by delicate perlitic partings. The rocks of the region are thus comparable with the lava flows of more recent age. Some of them were obsidian, others were lithoidal rhyolites and others holocrystalline rhyolites. The structure of the obsidians is now microcrystalline in consequence of the alteration or devitrification processes to which they have been subjected. They are now felsites or microgranites, but their microgranitic structure is not original. It is the result of devitrification. The author would therefor not call the rock a microgranite, nor an obsidian, but would designate it as an apobsidian or an aporhyolite, indicating that it was once an obsidian which has become devitrified—the preposition signifying that the rock to which it is prefixed has undergone alteration of a specific nature.

Another Occurrence of Websterite.—Another occurrence of the basic rock websterite is reported by Harker[1] from Fobello, Lombardy, Italy. The rock is a dark aggregate of black diallage moulding smaller grains of hypersthene. In thin section the diallage is colorless. An eclogite from Port Tana, Norway, consists of garnets holding inclusions of cyanite, omphacite and zircon, imbedded in a groundmass composed chiefly of colorless omphacite and quartz, in which lie phenocrysts of idiomorphic enstatite. A garnet amphibolite from Sutherland, England, a quartz diorite from Viti Leon, Figi, and a uralitized gabbro from Ena, Tonga Islands, are also described by the same author.

Petrographical News.—The nickel ores of Sudbury, Ontario, like those of Norway and Sweden, are associated with gabbro and norite, along their contact with other rocks. The ores are supposed by

[1]Geol. Magazine, VIII, 1891, p. 1.

Vogt[8] to be cencentrations from the magma that yielded the gabbro since the olivine of this rock often contains small percentages of nickel and other comparatively rare metals. The principal ore is a nickel marcasite with 3—5.5 per cent. Ni. The same author describes a nickeliferous pyrite from Beiern, Norway, whose density is 4.6, crystallization regular and hardness 4. It is not magnetic.

A peculiar quartz-porphyry consisting of quartz phenocrysts and crystals of apatite and an altered mineral supposed to be enstatite imbedded in a very fine grained weakly doubly refracting groundmass, which is water clear in thin section except where bespattered with dust inclusions or amorphous iron oxide, is mentioned by Hornung[9] as probably forming a sheet among the diabases and clay slates near Stalberg in the Harz.

Since many of the Maryland granites enclose fragments of other rocks that have suffered contact metamorphism, and since their microscopic constituents possess the characteristics of substances that have solidified from fusion, while the rock masses are intrusive in other rocks Keyes[10] believes he is justified in regarding them all as eruptive in origin. ·

Piedmontite from a new American Locality.—The rhyolites[11] of the South Mountain region in Pennsylvania and Maryland are characterized by their pink or bright red color, which, according to Williams,[12] is due to the large quantity of *piedmontite* in them. This rare manganese epidote occurs as a constituent in the rock mass, as radiating fibres filling veins and as well terminated crystals enclosed in *scheelite* occupying cavities in the rock. The latter were well enough developed to afford material for optical study. The elongation of the crystals is parallel to *b*. Their pleochroism is A=yellow; B= amethyst; C=carmine. Optically they are identical with piedmontite from other localities. An analysis gave (after correcting for quartz):

SiO_2	Al_2O_3	Ce_2O_3	R_2O_3	Fe_2O_3	Mn_2O_3	MnO	CaO	MgO	K_2O	Na_2O	H_2O	CuO	PbO	Total
37.37	22.07	.89	1.52	4.78	8.15	2.28	18.83	.30	.81	.27	2.48	.13	.17	=100.05

a result indicating that the South Mountain mineral is intermediate in composition between allanite, true piedmontite and mangan-epidote.

[8]Norges Geol. Undersög., 1892.
[9]Min. u. Petrog. Mitth., XIII, p. 373.
[10]Bull. Geol. Soc. Amer., Vol. IV, p. 299.
[11]AMERICAN NATURALIST, March, 1893, p. 273.
[12]Amer. Jour. Sci., 1893, XLVI, p. 50.

The mineral, when in the groundmass of the rhyolite is often associated with a pale rose epidote (*withamite*) and the common green variety, the latter in some cases surrounding the piedmontite. All of the epidotes are supposed to be of secondary origin.

Some American Minerals.—The interesting mineral *rowlandite* from Llano Co., Texas, to which reference has already been made in these notes, has recently been described by Hidden and Hillebrand[13]. Its color varies from bottle green to a pale drab green shade. It is more vitreous than gadolinite, is transparent in thin splinters and it weathers to a waxy brick red substance. The mineral is isotropic. Its hardness is 6 and its density 4.515. An analysis gave:

SiO_2	X	ThO_2	Ce_2O_3	La_2O_3 etc.	Yt_2O_3 etc.	Fe_2O_3	FeO	MnO	CaO
26.04	.39	.59	5.06	9.34	47.70	.09	4.39	.67	.50

MgO	Alk	H_2O	CO_2	Fl	P_2O_5	Total—O=F		
1.62	.28	.24	.34	3.87	tr	= 101.12—1.63 = 99.99.		

Disregarding the CO_2 and CaO and reducing the rare earths to a hypothetical one with the molecular weight of the yttrium group the formula becomes $Si_4 Yt_4 Fe Fl_2O_{14}$ or $Fe (YtF)_2 Yt_2 (Si_2O_7)_2$.

Transparent *xenotine* in small crystals associated with muscovite in a quartz pocket is reported by Hidden[14] from near Sulphur Spring, Alexander Co., N. C., and a green variety of the same mineral from the Brindletown gold district, Burke Co., in the same State. The green xenotine has been found only in the gold gravels, forming the interior portions of some of the rough brown crystals intermingled with the sand. It is thought to be original substance from which the brown material was derived by weathering. An analysis of the green mineral indicates a complicated composition:

SiO_2	ZrO	UO_2	ThO_2	Al_2O_3	Fe_2O_3	$(La Di)_2O_3$	$(Yt Er)_2O_3$	CaO	P_2O_5	F	H_2O
3.46	1.95	4.13	tr	.77	.65	.93	56.81	.21	30.31	.06	.57

In a paper entitled "Minerological Notes" Moses[15] describes *pyrite* crystals from a cavity in limestone at King's Bridge, N. Y. The crystals are octahedral in habit, with the octahedral faces striated parallel to $\infty O\infty$ and $\infty O2$. On the diploid and pyritoid faces the striations are parallel to their intersections, while the cubic faces are unstri-

[13]Amer. Jour. Sci., XLVI, 1893, p. 208. Cf. also Amer. Nat., 1893, p. 248.
[14]Ib., XLVI, 1893, p. 254.
[15]Ib., XLV, 1893, p. 488.

ated. The same author[16] has analyzed *ettringite* from the Lucky Cuss Mine, Tombstone, Arizona. The mineral is in aggregates of radiating fibres resembling in appearance a fibrous pectolite. These fibres are doubly refracting and have apparently a parallel extinction. The analysis of selected material gave:

SiO_2	Al_2O_3	CaO	SO_3	H_2O at 115°	Loss at red heat		Total
1.901	10.157	25.615	17.675	33.109	10.872	=	99.329

Reduced, these figures correspond with the formula

$$2\,[(H_{\frac{14}{5}}\,Ca\,Al_2)\,O_6]_5\,SO_3+8\,H_2O.$$

Pentlandite occurs at the Sudbury Mine in Ontario, intergrown with massive pyrrhotite. Penfield finds[17] its density to be about 5, and its composition: S =33.42; Fe =30.25; Ni = 34.23; Co = .85; gangue = .67. This corresponds to (Fe Ni) S, in which Fe: Ni = 1:1.32. The three supposed new sulphides *folgerite*, *blueite* and *whartonite* described by Emmens from this locality are thought by Penfield to be nickeliferous pyrite (blueite and whartonite) or mixtures of pentlandite with some impurity (folgerite).

Hidden reports[18] two new localities for gem *turquoise*. One is in the Cow Springs district of Grant Co., N. M., fifteen miles south of the Azure Mining Company's claim in the Burro Mountains, and the other is 150 miles east of the Burros in the Jarilla Mountains, Doña Ana Co., in the same State. Both localities were formerly worked by the natives. The matrix of the mineral in both cases is a trachyte traversed by fissures filled with quartz, limonite, kaolin, jarosite and other minerals. The kaolin is the result of alteration of the trachyte and the turquoise is regarded as a further alteration product of the kaolin.

A list of the minerals known to occur in Michigan is given by Hubbard.[19] Among these is a talc which the author calls *beaconite*. It occurs in fibres like those of asbestus, with an index of refraction = 1.5–1.6, an optical angle 2V = 60°, and a density of 2.74–2.88 Their composition as found by Packard is:

[16] Cf. also Zeits. f. Kryst., XXII, p. 16.
[17] Ib., XLV, 1893, p. 493.
[18] Ib., XLVI, 1893, p. 400.
[19] Rep. State (Mich.) Board of Geol. Survey, Lansing, 1893, p. 171.

SiO$_2$	Fe$_2$O$_3$. FeO	MnO	MgO	Ign	Total
59.72	8.67	.64	26.42	4.13 =	99.58

corresponding to H$_2$ (Mg Fe)$_3$ (SiO$_4$)$_3$.

A pink vitreous *zoisite* found at the Flat Rock Mine, Mitchell Co., N. C., associated with monazite and allanite, has been analyzed by Eakins.[20] Its composition is:

SiO$_2$	Al$_2$O$_3$	Fe$_2$O$_3$	MnO	CaO	H$_2$O	Total
38.98	31.02	4.15	.23	23.80	2.03 =	100.21

Specimens of *cacoxenite* from six localities have been examined optically by Luquer.[21] All the crystals show parallel extinction, and a few of the larger ones appear pleochroic in orange and light yellow tints. From a few measurements the approximate axial ratio 1: .75 was calculated.

The *heulandite*[22] from McDowell's quarry, Upper Montclair, N. J., crystallizes in forms agreeing essentially with those of crystals from Baltimore.

The material of the pale green crystals of muscovite from the dolomite of King's Bridge, N. Y., is a mica of the first order. Its apparent axial angle is 2E = 62° 11', 2E = 60° 37'.

Mineral Syntheses.—The ferrous bye-products of aniline factories at Laar, near Ruhort, Westphalia, when dumped upon the ground to dry, are so rapidly oxidized that 'the heaps soon become too hot to handle. The material hardens and assumes a metallic lustre.[23] On the walls of cavities within it crystals form whose habit is that of *hematite* but whose composition indicates an admixture of hematite with magnetite.

Upon heating to 1200° in a graphite crucible for several hours, one part of titanic iron and two and a half parts of pyrite, Michel[24] obtained a crystalline mass with the properties of *pyrrhotite*. This is filled with vacuoles on whose walls are implanted tiny crystals of *rutile* with the characteristics of the natural mineral.

Monticellite in well developed acicular crystals is reported by von

[20]Amer. Jour. Sci., 1893, XLVI, p. 154.
[21]Ib., 1893, XLVI, p. 154.
[22]A, J. Moses: School of Mines Quart., XIV, p. 326,
[23]Zeits. d. deutsch. geol. Ges., XLV, p. 63.
[24]Bull. Soc. Franc. d. Min., XVI, p. 37.

Gümbel[25] as existing in the slowly cooled silicate slags from the lead furnace at Friehung near Vilseck in Bavaria.

·V. Goldschmidt[26] calls attention to the advantage of glass over charcoal in securing sublimates of volatile substances arising during blowpipe analysis. He also gives the description of an apparatus which enables the manipulator to reduce his metallic compounds upon charcoal and collect their sublimates upon ordinary object glasses.

[25]Zeits. f. Kryst., XXII, p. 269.
[26]Zeits. f. Kryst., XXI, p. 329.

Reprinted from The American Naturalist, June 1st, 1894.

MINERALOGY.[1]

Contributions to Swedish Mineralogy, Part I:—In this paper Sjögren[2] has given in English a very interesting series of crystallographical studies. The well known but rare axinite from Nordmarken is reexamined. In addition to the tabular crystals described by Hisinger and v. Rath's prismatic type, a third type of smaller crystals is identified having neither the tabular nor the prismatic habits and highly modified. Hedyphane which is closely related chemically to the members of the apatite group, particularly mimetite, has been supposed to possess monoclinic symmetry on the basis of Des Cloiseaux's determination in 1881. Sjögren has examined crystals from the Harstigen mine in Wermland and finds that both crystallographically and optically hedyphane is hexagonal. The crystals examined exhibited the forms oP, ∞ P, P, $\frac{1}{3}$P, 2P, P2, 2P2, and clearly belong to the apatite group. Another member of the apatite group is discovered in Sjögren's new mineral svabite, which occurs in schefferite at the Harstigen mine. Svabite is a hydrous calcium arsenate of the composition indicated by the formula $HO Ca_5 As_3 O_{12}$ in which the hydroxyl appears to be part replaced by chlorine and fluorine. The mineral is crystallographically like apatite and exhibits the forms ∞ P, P, P2, oP. The same mineral was found at Jacobsberg, enclosed in hausmannite. A very exhaustive study is made of the minerals of the humite group, all of which are found at Nordmarken. No less than 29 forms were observed on chondrodite from this locality, and these include the six new forms, $+\frac{1}{3}$P,$-\frac{1}{3}$P,$+\frac{1}{3}$P,$-\frac{5}{2}$$\bar{P}\frac{5}{2}$,$+\bar{P}2$,$-\frac{1}{14}$$\bar{P}\frac{5}{2}$. The humite of the locality showed 20 and the clinohumite 26 forms, all of which have been observed on Vesuvian crystals. A probable fourth member of the humite group which occurs at Nordmarken, is announced in this paper. Three new analyses of longbanite are contributed, on the basis of which the formula of the mineral is given as $m Sb_2O_3 n Fe_2 O_3 p R^{IV}R''O_3$ in which R^{IV}=Mn and Si, and R''=Mn, Ca, and Mg. The symmetry of the mineral is shown to be rhombohedral, this and the chemical constitution indicating its isomorphous relation with hematite and ilmenite. Adelite is the name given to a new basic arseniate from Nordmarken, Jacobsberg and Longban, having the for-

[1]Edited by Dr. Wm. H. Hobbs, University of Wisconsin, Madison, Wis.
[2]Bull. of the Geol. Inst. of Upsala, I; No. 1, (1892), pp. 1–64, pls. I–IV.

mula 2CaO, 2MgO, H$_2$O, As$_2$O$_5$. The symmetry of the mineral is
monoclinic and its relationships, both chemical and crystallographical, are with triploidite, wagnerite and sarkinite.

Optical Methods :—Friedel[3] has devised a new method for determining the double refraction in thin sections of minerals on the stage
of the ordinary petrographical microscope.   The method makes use of
the quarter undulation mica plate.   The nicols are crossed and the
slide is raised a short distance above the stage on thin blocks, so as to
allow of the introduction of the mica plate between the slide and the
stage.   The stage is now revolved until the directions of extinction
make 45° with the principal sections of the nicols.   The mica plate is
introduced below the slide and carefully turned without moving the
stage until that portion lying outside the mineral plate is extinguished.
By now revolving the polarizer, the mineral can be extinguished or
given the same illumination as the mica plate.   The observations are
made in monochromatic light.   If the positive direction of the mineral
plate passes through the upper right quadrant of the field and the
positive direction of the mica plate coincides with the vertical cross
hair, the polarizer should be revolved to the right, the angle φ required
to produce extinction, and the angle φ_1 required to produce equal
illumination of mineral plate and mica plate, yielding ψ the difference
in phase produced in the mineral section.   The formulas are $\psi=\varphi_1$
and $\psi=2\varphi_1$.   The greater part of the paper is devoted to methods of
evaluating errors in the process.

Harker[4] has determined trigonometrically the values of the extinction angle in prismatic cleavage flakes of augite and hornblende, as
dependent on the optical angle and the extinction angle in the plane
of symmetry.   His tables of values will be convenient for reference,
but as he points out, the variation in the values with 2V is not great
enough to determine the optical angle from measurements of the prismatic and clinopinacoidal extinction angles.

Isotypism :—Rinne[5] compares crystals of the metals with crystals
of their oxides, sulphides, hydroxides and haloid compounds.   He
points out that in this comparison we find strikingly close relationships between bodies markedly different chemically, and these relationships do not consist simply in identity of crystal symmetry, but in

[3]Bull. Soc. Franç. Minér., XVI; 19 (1893).
[4]Min. Mag., X (No. 47), p. 239.
[5]Neues Jahrb. f. Min., etc., 1894, (I) pp. 1–55.

close approximation to a type as regards crystal shape (Krystallgestalt) and interfacial angles. Even when the symmetry of two substances is not identical, he makes comparison of the crystal shape as, e. g., between a cube and a rhombohedron with polar edge approaching 90°. The author distinguishes seven types as follows: I regular type (isometric), II magnesium type (hexagonal and pseudo-hexagonal—orthorhombic), III arsenic type (rhombohedral), IV quartz type (hexagonal tetarto-hedral), V $\widehat{a}$ tin type (tetragonal), VI rutile type (tetragonal and pseudo-tetragonal—orthorhombic), VII β tin type (orthorhombic and pseudo-orthorhombic—monoclinic). Every group but the fourth contains metals and this type Rinne considers as derivable from the third or arsenic type. Many oxides, etc., have their crystal forms to some extent indicated in the forms of their contained metals. The term isotypism is proposed to describe these crystallographical relations between members of different divisions of the chemical mineral system. The author further states, " It must now be accepted as a fact that such substances " (elements, oxides, sulphides, haloid salts, and even silicates, which have been grouped together under his various types) " possess equivalent or very similar crystal forms, and it follows that the chemical differentiation into elements, oxides, salts, etc., finds no crystallographical expression, and therefore no independent, certain conclusion as to the chemical group to which a compound belongs can be drawn from its crystal form."

Lamellar Structure in Quartz Crystals.—In an "additional note on the lamellar structure of quartz crystals and the methods by which it is developed," Professor Judd[6] describes and figures a remarkably beautiful instance of lamellar structure in quartz, in which he sees a close analogy with the "rippled fracture" which he finds can be produced in quartz crystals by breaking them in a powerful vice along a plane perpendicular to the optic axis. The appearance of such fractures is very much like that of "engine-turned surfaces." This appearance is caused by ridges following the planes R and –R, which are often curved and die out in the manner of plagioclase lamellæ. From a study of the lamellæ in an equatorial section of quartz supposed to be one of those investigated by Brewster, Professor Judd concludes that quartz is dimorphous. What he calls "stable quartz" shows no tendency to assume a lamellar structure, whereas "unstable quartz" constantly exhibits such a tendency. The latter variety is usually amethystine. The lamellæ consist of alternating bands of

[7]Min. Mag., X, p. 123.

right and left handed quartz.   When they are bent or disturbed they
furnish biaxial interference figures.   Many crystals are composed of
both stable and unstable quartz, the relative positions of which show
some relation to the symmetry of the crystal.   Such crystals, or crystals composed entirely of unstable quartz, have the lamellæ induced
by great mechanical stresses.   The fact that the structure is only
faintly induced and that very near the fracture in artificially crushed
crystals, is explained by the short time during which the stress is
applied, permanent structure being produced only after a long application of the stress.

Reprinted from The American Naturalist, July 1st, 1894.

Contact Effects around Saxon Granites.—The effects of the granite and syenite of Lausitz, of the granitite of Markersbach and of the tourmaline granite of Gottleube upon the rocks through which they cut in the Elbthalgebirge in Saxony, are concisely described by Beck.[2] The members of the phyllite formation and the beds of Cambrian, Silurian and Devonian age, whatever may have been their nature, have all undergone contact metamorphisen near ther junction with the eruptives. During the process of alteration there seems to have been little addition of material to the metamorphosed rocks, as all the contact products when originating from the same member of the bedded series are the same, irrespective of the nature of the metamorphising eruptive. The great variety in the contact products of the region is due solely to differences in the character of the originals of the altered rocks. The phyllites have been changed to 'Fruchtshiefer' and into andalusite mica schists, chlorite gneisses into biotite gneiss, and feldspathic quartzites into hornfels. The Silurian slates near the contacts have become hornstones and knotty schists, carbonaceous quartz schists have changed into graphitic quartzites, graywackes and marbles have been made crystalline, and the latter rock has in many cases been changed into a calc-silicate aggregate, which has been impregnated with ore masses, presumably originally in the granitite with which the limestones were in contact. Diabases and diabase tuffs in proximity to the intrusive rocks have been amphibolized. The Devonian rocks have suffered the same alterations as the corresponding Silurian ones, and in addition there has been formed a gneiss-like rock whose predecessor among the clastics is unknown. A large number of contact minerals are discussed at length by the author, chief among them being quartz, plagioclase, cardierite and graphite. The article is full of instructive suggestions though nothing of striking novelty is met with in it.

The Schists of the Malvern Hills.—Callaway[3] has published a final summary of the conclusions based on seven years work in the Malvern Hills. He reiterates his belief that the schists of the region

---

[1]Edited by Dr. W. S. Bayley, Colby University, Waterville. Me.
[2]Min. u. Petrog. Mitth. XIII, p. 290.
[3]Quart. Jour. Geol. Soc., XLIX, p. 398.

are squeezed eruptives, and discusses the physical, mineralogical and chemical changes that have effected the alteration of the granites and diorites into gneisses and schists of various kinds. His conclusion that a sericite schist may be derived from diorite and that biotite is often an alteration product of chlorite are both of great interest. In the change of a massive into a schistise rock, the author states that the former "passes through the intermediate state of a laminated grit, which thus simulates a true sediment, the subsequent stages of alteration and cementation resembling the process of metamorphism in some bedded rocks." In the production of the foliation there is decomposition of the original components of the massive rock and a reconstruction of new minerals largely from these decomposition products. In the Malvern Hill rocks orthoclase has been replaced by quartz and muscovite, plagioclase by quartz and muscovite, chlorite by biotite and white mica, and biotite by a white mica. A number of analyses appear in the paper to illustrate the chemical changes that have accompanied the physical ones through which the respective rocks have passed.

A Soda-Rhyolite from the Berkeley Hills, Cal.—In the Contra Costa Hills near Berkeley, California, are occurrences of a volcanic flow that has been investigated by Palache,[4] who recognizes three facies of the rock. In the first, the porphyritic phase, phenocrysts of quartz and feldspar are abundantly disseminated through a micro-granular aggregate of the same minerals. The second phase is characterized by the possession of numerous small spherulites in a glassy matrix, in which are a few small grains of magnetite and some feathery aggregates of chalcedony. The third phase is a glass containing tiny microlites of feldspar and grains of magnetite. Analyses of the different types indicate that the material of each type has the composition of a soda-rhyolite. The spherulitic variety which is intermediate between the other two, in its acidity is composed as follows:

SiO_2	Al_2O_3	Fe_2O_3	CaO	MgO	K_2O	Na_2O	H_2O	Total	Density
75.46	13.18	.91	.95	.10	1.09	6.88	.93	= 99.50	2.42

Diabases from Rio Janeiro, Brazil.—Sections from a series of twelve diabase dykes from Rio Janeiro, Brazil, have been investigated by Hovey,[5] with some interesting results. The chemical composition

[4]Bull. Dept. Geol. Univ. Cal., Vol. 1, p. 61.
[5]Min. u. Petrog, Mitth. XIII, p. 211.

of all the dykes is practically the same. Their mineral composition
and structure, however, vary. In the largest dykes the number of
constituents discovered is much greater than in the smaller ones. They
embrace the usual diabase components with the addition of a light colored sahlitic pyroxene differing from the sahlite of Sala in the value
of its optical angle. In the Brazil mineral $E_a = 32° 39'$, while in the
Sala mineral it is 112° 30'. It is the oldest constituent of the rock
after magnetite, and, consequently it is that which approaches most
nearly to being idiomorphic. The structure of the large dykes is
gabbroitic and ophitic, whereas that of the small ones is porphyritic
and hyalopilitic, with the pyroxene figuring as the phenocrysts.
Quartz is not uncommon in the coarser rocks and granophyric intergrowths of quartz and feldspar are frequently met with.

The New Island off Pantelleria—A Correction.—In these
notes for December[6] last, the statement was made concerning the
material of a recent eruption near Pantelleria, that it consisted of loose
blocks and of lava. Mr. G. W. Butler of Chertsey, England, corrects this statement in a recent letter to the writer and declares that the
new island formed during the eruption was composed entirely of loose
scoriaceous bombs, which disappeared a short time after the eruption
ceased.[7]

Petrographical Provinces.—Iddings[8] gives a brief and, consequently, a tantalizing account of the old volcano of Crandall Basin in
the Absaranka Range of Mountains in the Yellowstone National Park,
that has been eroded in a manner to give a good section of the cone
with the dykes and flows radiating from it. The different rock types
mentioned in the paper are simply alluded to, a full account of them
being promised later. The author's conclusion from his study is to the
effect that we have here proof that the texture of rocks and their mineral composition is more directly dependent upon the rapidity with
which the rocks cooled, than upon the pressure to which they were
subjected during their solidification. The differentiation of rock magmas is also well shown in the case of the volcano studied by the production of many individual rock types.

Upon comparing thirty-nine of the best analyses of rocks occurring
in the eruptive areas around the Bay of Naples, Lang[9] concludes that

[6]AMERICAN NATURALIST, Dec., 1893, p. 1088.
[7]Cf. also G. W. Butler; *Nature*, April 21, 1892.
[8]Jour. Geol., Vol. 1, p. 606.
[9]Zeits. d. deutsch. geol. Ges. XLV, p. 177.

there are here three independent volcanic centers, represented respectively by Ischia, Vesuvius and Mt. Nuovo. That they are on different volcanic fissures is indicated by the differences in the character of the lavas extruded from them, especially in their sodium and calcium contents. At each center each magma became differentiated, and this differentiation explains the variety of the rock types discovered in each.

'A study in the consanguinity of eruptive rocks' is the title of an article by Derby[10] in which is shown the fact that the occurrence of the eleolite syenites, phonolites, monchiquites and other related rocks in Brazil, point to the correctness of the notions of differentiation and consanguinity as explanatory of the existence of different phases of eruptive rocks within the same volcanic sphere. The author also shows that, while not having formulated the theory, its principle has been the guide in his work on the Brazilian rocks.

Miscellaneous.—Upon examining spherulites of lithium phosphate between crossed nicols, McMahon[11] finds that some of the groupings present apparently miaxial crosses which remain fixed in position during a complete revolution, while in others the cross breaks up into two hyperbolic branches resembling those of biaxial optical figures. The phenomenon, the author regards as due to molecular strains that affected the spherulites at the time of their crystallization.

[10]Jour. Geol., Vol. 1, p. 579.
[11]Mineralogical Magazine, X, p. 229.

Reprinted from The American Naturalist, July, 1894.

MINERALOGY.[1]

Friedel's Cours de Mineralogie.[2]—The first part of a text-book of mineralogy by Charles Friedel covers the field of general mineralogy. In the preface it is stated that a second part, devoted to special or descriptive mineralogy, will be prepared with the assistance of M. George Friedel, the author's son. The book does not claim to be, the author states, a treatise on crystallography or crystal physics, but a practical method of determining minerals on the basis of their morphological, physical, and chemical properties. It is intended for the use of those students who are preparing for the examinations for licentiate in physical sciences, and should therefore be adapted to the needs of college students.

The book contains 416 pages with the subject matter distributed as follows: introduction (giving history of science and fundamental definitions, 16 pages); organoleptic properties, 16 pages; crystallography, 238 pages; physical (and optical) properties, 59 pages; chemical composition occupies the remainder of the book and includes the divisions, blowpipe methods, mineral synthesis, and mineral classification. Under organoleptic properties are included among others, structure, color, lustre, density, external form (with a consideration of pseudomorphs), hardness, and streak. In treating crystallography eight pages are devoted to an exposition of Hauy's *théorie des décroissements*. This is followed by sections on the law of rational indices and symmetry. After deriving the crystal systems, the author gives eight pages to an exposition of Bravais's theory of crystal structure. No mention is made of the work of later writers on this subject, and throughout the book a tendency to utilize mainly the work of French writers seems manifest. The difficulties of translating Levy's symbols into those of Weiss, Naumann, Dana and Miller, makes it necessary to devote thirty-seven pages to crystallographic notation. Twelve of these are consumed by a table giving the equivalents of Levy's symbols in the other notations. An usually large amount of space for a book of this sort is devoted to the representation of crytals, but those which illustrate the book are very poor. Many of the figures are not merely carelessly, but incorrectly drawn. Crystals having a principal

[1] Edited by Dr. Wm. H. Hobbs, University of Wisconsin, Madison, Wis.
[2] Cours de minéralogie professé a la faculté des sciences de Paris, par Charles Friedel. Minéralogie générale, pp. iii and 416. Paris, 1893.

axis are generally lopsided. Figures 70, 138, 224, 255 and 322 are a few of the incorrectly drawn crystals. Another bad feature of the illustrations is that crystals are not always properly set up but are seen from all directions. The best portions of the work are those which treat optical mineralogy and mineral synthesis. The former is treated without mathematics and in a simple and practical manner. The section on the classification of minerals is very unsatisfactory. What .purports to be a history of the subject is given. The systems mentioned are those of Werner, Hauy, Beudant, Delafosse and Dana. Groth's system is not mentioned nor is that of any other modern German mineralogist. A considerable number of pages is devoted to detailed lists of minerals as they appear in the schemes of Werner, Delafosse, and Dana. With the exception of the latter, which Friedel adopts as the one most in harmony with the present state of the science, these lists seem out of place. The book is not provided with an index, but has a somewhat extended table of contents.

As a text-book the work is subject to criticism on account of its classification and arrangement of subject matter, its lack of perspective in the treatment of the different divisions of the subject, its tendency to utilize mainly French investigations and systems, and its faulty illustrations.

Relation between Atomic Weight and Crystal Angles.—

In a paper entitled, "Connection between the Atomic Weight of contained metals and the magnitude of the angles of crytals of isomorphous series, a study of the potassium, rubidium and cæsium salts of the monoclinic series of double sulphates $R_2M (SO_4)_2$ 6 H_2O," Tutton[1] has given the results of a most careful and thorough crystallographical study of an isomorphous series of salts, to determine the kind and degree of effect which the different bases exert upon the crystal angles. The results are very interesting since they seem to show a relation between the atomic weights of the contained bases and the crystal angles. The work involved no less than 9,500 measurements. The crystals were obtained by slow crystallization from cold solutions and ten good crystals of each salt were selected for measurement from a dozen or more different crops. The double salts of the formula R_2M $(SO_4)_2$ 6 H_2O containing as univalent metals either potassium, rubidium, or cæsium, and as bivalent metals either magnesium, zinc, iron, manganese, nickel, cobalt, copper, or cadmium, were always pre-

[1] Jour. Chem. Soc. London, Trans., Vol. LXIII, (1893), pp. 337–423.

46

pared by mixing solutions of the two simple sulphates in equal molecular proportions. The study shows that the bivalent metal exerts no appreciable effect on the crystals, the predominant effect being due to the univalent metal present. The crystals of the potassium, rubidium, and cæsium salts have each a peculiar habit, that of the rubidium being intermediate between the other two. The axial angle β increases from the cæsium, through the rubidium to the potassium salt, its value in the rubidium salt being midway between the values in the cæsium and potassium salts. This is in close correspondence with the differences between the atomic weights of those bases. Tutton says "The relative amounts of change brought about in the magnitude of the axial angle by replacing the alkali metal potassium by rubidium and the rubidium subsequently by cæsium, are approximately in direct simple proportion to the relative differences between the atomic weights of the metals interchanged." The other crystal angles of the rubidium salts are likewise intermediate in value between those of the potassium and cæsium salts, but they do not show the same relation to the atomic weights of the alkali bases, the maximum deviation from such a relation being found in the prism zone. As these angles are for rubidium nearer to those of potassium than to those of cæsium, the author thinks that as the atomic weight of the alkali metal introduced gets higher, the effect of the metal on certain angles increases beyond a mere numerical proportion. Professor Tutton announces that this communication will be followed by another, which will discuss the changes in the optical constants of the crystals due to the same chemical substitutions.

Spangolite from Cornwall.—Miers[4] has found in a collection of Cornwall minerals presented to the British Museum, small crystals of the new mineral spangolite described by Penfield in 1890. The Cornwall crystals show the hexagonal prism, pyramid, and base. Their association is remarkably like that of Penfield's spangolite, as they occur with cuprite and its alteration products. From the characters of the associated liroconite and clinoclase, Miers thinks that there can be no doubt that the specimen is from St. Day, near Redruth.

Eudialite from the Kola Peninsula.—The occurrence of eudialite in the nephelene syenite and pegmatite of the Lujawr-Urt and Umptek in Russian Lapland, recently mentioned by Ramsay, has now been studied in detail.[5] The crystals have developed on them the

[4] Neues Jahrbuch, 1893, II, 174.
[5] Neues Jahrbuch, Beil. Bd., VIII, (1893) 722.

forms R,—$\frac{1}{2}$R, $\frac{1}{2}$R,—2R, ∞ R2, ∞R, and oR. The axial ratio is a:c =1:2.1072. The mineral has good cleavage parallel to the base and one varying from very good to poor runs parallel to the second order prism. The color is usually cherry to garnet red. The crystals are specially interesting because of a marked zonal structure and of a division into sectors having differences in double refraction. Some of these sectors have positive and others negative double refraction. Like the eudialite from Magnet Cove the crystals are optically anomalous, sometimes having biaxial character with optical angle as large as 15°. On heating the sections of the crystals to a temperature at which boracite had become isotropic, all the sectors of the field seemed to give negative double refraction. Ramsay finds evidence that the different zones of the mineral possess different specific gravities as well as different double refraction, and he considers this to be due to isomorphous growth together of eudialite and eucolite. He shows that as regards axial ratio, specific gravity, double refraction and optical character, there is a gradation from the eucolite of Arö through the eudialites of Umptek and Kangerdluarsuk to the eudialite of Magnet Cove.

Reprinted from The American Naturalist, August 1st, 1894.

PETROGRAPY.[1]

The Ejected Blocks of Monte Somma.—Johnston Lavis[2] has begun a thorough study of the ejected blocks of Monte Somma, with especial reference to their petrography and the nature of the metamorphic changes that have been produced in them by the lavas by which they were enclosed. The druse minerals of the blocks have long been known, but their nature as rocks has been left uninvestigated. The author proposes to study in detail about 700 specimens of the blocks, including many varieties. He begins by describing some 30 that were originally stratified Cretaceous limestones containing carbonaceous material. The first stage in their alteration seems to be the conversion of bituminous substance into graphite, and the crystallization of the rock into marble. The crystallization has not destroyed the original bedding bands, nor the most delicate structures exhibited by them, hence it is assumed that fusion or softening of the rock did not accompany the crystallization processes. A few olivines were formed at this time, and these consequently are the first products of the metamorphosing agency. They appear principally as inclusions in the calcite. In the next stage of alteration the graphite disappears, and a saccharoidal marble results. This contains more or less colorless olivine, and passes rapidly into a mass of olivine, colorless pyroxene, wollastonite and biotite, where impurities were present in the original rock. In the earlier stages of metamorphism the calcite and the silicate minerals will exist in different bands, but in later stages silicates and calcite intermingle, and finally a purely silicate rock results. The order in which new minerals seem to develope is thought to be the following; olivine, periclase, humite, spinel, mica, fluorite, galena, pyrite, wollastonite, garnet, vesuvianite, nepheline, sodalite, feldspar, secondary calcite, tremolite, brucite. The article is illustrated by three lithographic plates. It will repay close study by students of contact action, as we have recorded in the blocks the effects of the action of a magma upon a limestone, in all its stages.

Phonolites from the Black Hills.—The sanidine-trachyte described by Caswell[3] from Bear Lodge in the Black Hills, has been

[1] Edited by Dr. W. S. Bayley, Colby University, Waterville, Me.
[2] Tarns. Edin. Geol. Soc., VI, 1893, p. 314.
[3] U. S. Geog. & Geol. Survey of Rocky Mts. 1880. Cap. VII, p. 471.

reexamined by Pirsson,[4] who finds it to be a phonolite with phenocrysts of anorthoclase and pyroxene, in a groundmass of the usual components of phonolite. The anorthoclase has the composition:

SiO_2	Al_2O_3	Fe_2O_3	CaO	Na_2O	K_2O	H_2O	Total	Sp. Gr.
66.44	19.12	.56	tr	7.91	5.10	.57	= 99.70	2.585

The nepheline is all in the groundmass where it appears as idiomorphic crystals. The density of the rock is 2.582 and its composition:

O_2	TiO	Al_2O_3	Fe_2O_3	FeO	MnO	CaO	BaO	MgO	Na_2O	K_2O	H_2O	Cl	SO_3	Total	Cl
.08	.18	18.71	1.91	.63	tr	1.58	.05	.08	8.68	4.63	2.21	.12	tr	= 99.86	—.08 = 99.88

A second occurrence of phonolite within the same region is in a dyke just south of Deadwood. It consists of phenocrysts of reddish feldspars and black hornblendes that approach barkevikite in properties. The rock from the Black Hills sold by the dealers as tinguaite is a dense aggregate of pyroxene phenocrysts in a matrix of feldspar and aegirine, with an occasional patch of nepheline.

The Origin of Norwegian Iron Ores.—The iron and other ores of many of the Norwegian localities are connected genetically with granites and gabbroitic eruptives. The iron ores in veins are supposed by Vogt[5] to be due to contact action between granite and the surrounding rocks. Those connected with the gabbros are basic accumulations, whose origin is ascribed to differentiation of the basic magma. In consequence of this differentiation, which is governed largely by Soret's principle and the differences in density of the various differentiated products, the gabbro splits into labrador-rock and various iron-olivine and iron-pyroxene compounds, and in these latter are accumulations of magnetite and ilmenite large enough to constitute ore bodies. Each of the iron-pyroxene rocks is described by the author and the iron ores associated with them are characterized. The titanium of the iron is thought to have originated mainly in the olivine and other basic components of the normal gabbro.

The Tonalites of the Rieseferner.—The tonalites of the Rieseferner in the Tyrol are again the subject of careful petrographical study.[6] The normal tonalite (hornblende–mica–quartz–diorite)

[4]Amer. Jour. Sci., XLVII, 1894, p. 341.
[5]Geol. Fören Stock. Förh. 13 and 14.
[6]Becke: Min. u. Petrog. Mitth., XIII, p. 379.

which is a coarse granular rock, on its periphery often becomes finer grained and porphyritic. Large biotites and hornblendes are scattered through its groundmass, which remains fine grained, and the rock thus takes on a prophyritic habit. At other times the decrease in the size of its constituent grains is accompanied by a decrease in the proportion of plagioclase and quartz present in the rock and a large increase in the orthoclase present, while hornblende disappears completely. It is unnecessary to give the petrographical details of the author's paper. It should be mentioned, however, that the feldspars are very carefully studied by comparing the differences in their refracting indices, and many new points are brought out concerning their relations to each other. Some of the plagioclases were found to consist of nuclei of basic plagioclase, enclosing areas of a more acid feldspar identical with an acid peripheral zone. The phenomenon is thought to be due to corrosive influences. In addition to the various phases of the tonalite mentioned, the author makes a careful study of the veins cutting them, and of the slight alterations they have suffered and he refers to the existence of gneiss fragments occasionlly met with in their peripheral portions.

Petrographical News.—McMahon' cites, as evidence in favor of the eruptive character of the Dartmoor granite, and in opposition to the view of Ussher that it resulted from the fusion by pressure of pre-existing pre-Devonian sedimentaries, the following facts. Its apophyses cut the surrounding rocks. The metamorphic changes effected in the latter are the result of contact action. Finally the other rocks with which the granite is associated show no evidence of the great pressure, to which they must have been subjected if the granite were truly a fused sedimentary.

Associated with the argillites, graywackes and other sedimentary rocks of the Keewatin series near Kekaquabic Lake in Northeastern Minnesota, Grant* has discovered volcanic fragmentals and amphibole schists, the former of which are recognized as diabase tuffs and the latter as their recrystallized representatives.

A quartz bearing leukophyre variety of diabase porphyrite, forms intrusive layers in the Carboniferous schists at the Hernitz Mine near Saarbrücken in the Pfalz.' The rock was regarded by Weiss as a melaphyre.

'Quart. Journ. Geol. Soc., XLIX, p. 385.
Proc. Somerset Arch. & Nat. Hist. Soc., Vol. 28, p. 892.
*Science, XXIII, 1894, p. 17.
'Laspeyres : Corr. Blatt. Naturh. Ver. Bonn., 1893, p. 47.

The tuffs found with the nepheline leucite basalts of the Dauner region in the Eifel are made up of augite, mica, and olivine fragments, augite crystals, glass particles and lapilli cemented together by quartz and felspar which represent an original glassy cement.[10]

On the west coast of the Island of Celebes, Wichmann[11] finds boulders of an epidote glaucophane-mica schist, supposed to be associated somewhere in the interior of the island with mica quartzite.

[10]L. Schulte: Verh. d. Naturh. Ver. Bonn., 1893, p. 295.
[11]Neues Jahrb. f. Min. etc., 1893, II, p. 176.

Reprinted from The American Naturalist, August 1st, 1894.

PETROGRAPY.[1]

In a long and extensive article, Mügge[2] treats of the keratophyres of the Lennethal in Westphalia, and the neighboring regions, and their tuffs. The rocks have been considered as fragmental schists by some observers and as squeezed eruptives by others. They are known generally as the Lenneporphyries. Mügge finds that some of them are genuine eruptives and some are the tuffs of these. The massive rocks are keratophyres and quartz-keratophyres, sometimes carrying large phenocrysts of quartz and feldspar and at other times free from these. The groundmass of the keratophyres is made up of bleached biotite, sericite, feldspar, opal and glass, with traces of spherulitic structure. Schistose varieties of the quartzose varieties have become foliated through pressure, as shown by the fractured quartzes and feldspars that occur so abundantly in them, the presence of lenticular areas of quartz mosaic and the greater abundance of sericite. The most characteristic of the lenneporphyries are tuffs in which the ash structure is very well exhibit. The typical tuff structure is described by the author as due to the accumulation of glass particles with concave boundaries. These are mingled with complete and broken crystals of various minerals and often with sedimentary material. Rocks composed of intermingled volcanic and sedimentary fragmental material the author would call tuffites; when metamorphosed, tuffoids. Many of the rocks in the Lenne district have suffered dynamic metamorphism with the production of secondary quartz, feldspar, sericite, carbonates and chlorite. They are, therefore, tuffoids. The new material was formed partially from the decomposition of the rock's materials and partially with the aid of alkaline solutions originating outside of the metamorphised rocks.

Nepheline-Melilite Rocks of Texas.—Osann[3] finds a melilite nepheline basalt occurring as dykes in the Cretaceous of Uvalde Co., Texas, and nepheline basanites forming buttes and hills in the same region. The basalts are typical melilite varieties, containing phenocrysts of olivine and micro-porphyritic crystals of melilite with all the characteristic features of this mineral. Perofskite is a common

[1] Edited by Dr. W. S. Bayley, Colby University, Waterville. Me.
[2] Neues Jahrb. f. Min., etc. B. B. viii, p. 525.
[3] Jour. Geol, Vol. I, p. 341.

accompaniment of the melilite. The basanites have an andesitic habit and since they contain more or less sanidine, they approach phonolite in composition. Hornblendes, two monoclinic augites and nepheline are common as phenocrysts, while sanidine, plagioclase and olivine are scarce. The rock of Pilot Knob, near Austin, is a porphyritic nepheline basalt.

Eleolite Syenite from Eastern Ontario.—Adams,[4] while making a geological reconnaissance in the township of Dungannon, Ontario, discovered a large area of eleolite syenite in the Laurentian of the region. The rock is notable especially for the fresh scapolite and calcite present in it and for the fact that its feldspathic constituent is an albite. Petrographically the syenite is an aggregate of the minerals above mentioned and hornblende, biotite, sodalite, garnet and zircon. The nepheline is fresh. It occurs in large quantity, and sometimes in individuals two and a half feet in length. Its composition according to Harrington is

SiO_2	Al_2O_3	Fe_2O_3	CaO	MgO	K_2O	Na_2O	Loss	Total
43.51	33.78	.15	.16	tr	5.40	16.94	.40 =	100.34

The mica is a dark yellow-brown variety. It is present in small quantities only. Hornblende is also comparatively rare. It occurs in two varieties in different specimens. One variety has a large optical angle and a pleochroism of deep green and pale yellow tints. The other is allied to arfvedsonite. It has a small axial angle, and is pleochroic in deep bluish-green and yellowish-green tints. The scapolite is in large colorless grains that are fresh and seem to be original, and the calcite in more or less rounded individuals, often included within the other constituents. The feldspar is largely albite. A small quanitity orothoclase occurs, especially associated with the sodalite. This orothoclase is thought to be secondary.[5] An analysis of the sodalite gave:

SiO_2	Al_2O_3	FeO	Na_2O	K_2O	Cl	SO_3	H_2O	Ins.	Total
36.58	31.05	.20	24.81	.79	6.88	.12	.27	.80 =	101.50

O = Cl 1.55 = 99.95.

Petrographical News.—The basic dyke material at Hamburg, Sussex Co., N. J., which was thought to be leucite tephrite by Hus-

[4] Amer. Jour. Sci., 1894, XLVIII, p. 10.

[5] Cf. also Geol. Surv. of Can., Vol. VI, Pt. J.

sak[6] and declared by Kemp[7] to be an aggregate of pyroxene, biotite and analcite has been examined at another place by the last named geologist. It has been found by him[8] to contain leucite. Hussak's determination is thus confirmed. The rock is a leucite tephrite.

A spherical granite from a boulder discovered on Qonochontogue Beach in Southwestern Rhode Island is described by Kemp[9] as a coarse granitite, with nodules from two to three inches in diameter scattered through it. These consist of a center of coarse plagioclase with a little quartz, surrounded by a concentric zone of biotite and magnetite, and a peripheral one of radiating plagioclase, whose laths end sharply against the granite matrix. The author explains the nodules as centers of crystallization.

The rocks that have for the past few years been called muscovadite by the Minnesota Geological Survey have recently been examined by Grant,[10] who finds among them several distinct rock types. Some of muscovadites are fine grained aggregates of pyroxene, quartz and feldspar, containing in their midst large flakes of biotite. Others are composed of quartz and biotite, etc. These are considered as contact rocks. A second class of the muscovadite comprises granulitic gabbros and norites.

The siliceous oolite of State College, Pa., is composed of radial spherules of fibrous chalcedony forming bands around fragments and rounded grains of quartz. Between the spherules are bundles of chalcedony fibres placed normal to the surface of the spherules nearest them, and intermingled with these are granular chalcedony and quartz. An oolite from the Tertiary beds of New Jersey is an aggregate of sphero-crystals of chalcedony, usually without nuclei. Occasionally a cone of fine grained quartz is to be seen, but this is rare. The matrix between the spherules is partly chalcedony and partly quartz.[11]

Duparc and Mrazec[12] refer very briefly to the mineralogical composition of an occurrence of Serpentine at Geisspfad in the Swiss Alps. The rock now contains hornblende, chromiferous diopside, diallage and some secondary substances in addition to serpentine. The rock was probably originally a Lherzolite.

[6]Amer. Naturalist, 1893, p. 274.
[7]Ib. 1893, p. 563.
[8]Amer. Jour. Sci , XLVII, 1894, p. 333.
[9]Trans. N. Y. Acad. Sci., XIII, 1894, p. 140.
[10]21st Ann. Rep. Minn. Survey, p. 147.
[11]E. O. Hovey: Bull. Geol. Soc. Amer., Vol. 5, p. 627.
[12]Bull. Soc. Franc d. Min., XVI, p. 210.

Phillips[13] has analyzed specimens of Pele's hair (I) and of lava stalagmites (II) from the caves of Kilauea, Hawaii, with these results:

SiO_2	Al_2O_3	Fe_2O_3	FeO	MnO	P_2O_5	CaO	MgO	Na_2O	K_2O	Total
50.76	14.75	2.89	9.85	.41	.26	11.05	6.54	2.70	.88	= 100.09
51.77	15.66	8.46	6.54	.82		9.56	4.95	2.17	.96	= 100.89

Lacroix[14] finds specimens of nepheline basalt from Saint Sandoux, Puy-de-Dom, France, in an old collection preserved in the College of France.

Some of the trap dykes of the Lake Champlain region are camptonites. Others consist of monchiquite, fourchite or bostonite. All are described by Kemp and Marsters[15] in a recent Bulletin of the Survey

[13]Amer. Jour. Sci., XLVII, p. 473.
[14]Bull. Soc. Franc d. Min., XVII, p. 43.
[15]Bull. U. S. Geol. Surv., No. 107.

Reprinted from The American Naturalist, September 1st, 1894.

General Notes.

MINERALOGY.[1]

Crystallization of Enargite.—Pirsson[2] has studied enargite from two new Colorado localities, viz., the Ida Mine, Summit District, and the National Belle Mine, Red Mountain. At the former locality the mineral is deposited in cavities left after the kaolinization of feldspar phenocrysts in porphyry. These crystals are tabular parallel to $\infty\,\mathrm{P}\overline{\infty}$, and are bounded by the forms $\infty\,\mathrm{P}\overline{\infty}$, oP, ∞P, and $\infty\,\mathrm{P}2$. At the latter locality two types of crystals are found. One of these is in thick, striated prisms bounded by the same forms as the Ida Mine crystals and sometimes in addition $\mathrm{P}\widetilde{\infty}$, $\mathrm{P}\infty$, $\infty\,\mathrm{P}\check{}$, and another brachydome. The second type of crystals from this locality is tabular parallel to the base and shows hemimorphic development. The forms observed on this type are oP, $\infty\,\mathrm{P}\infty$, $\alpha\mathrm{P}$, $\infty\mathrm{P}3$, $\mathrm{P}\widetilde{\infty}$, $\tfrac{3}{2}\,\mathrm{P}3$.

Crystallization of Scolecite and Meta-scolecite.—Rinne[3] has investigated crystals of scolecite from Iceland and shown that the mineral crystallizes in the rare inclined-faced hemihedral division of the monoclinic system. This fact was developed by etching and by study of the pyroelectric properties. The front faces of the prism have different etched figures from the rear faces, while in twinned crystals with the twinning plane the ortho-pinacoid, front and rear faces of the prism have the same figures. In simple individuals the front and rear faces are pyroelectrically positive and negative poles respectively. In twinned crystals all prism faces are positive and a negative zone follows the twinning line on $\infty\,\mathrm{P}\widetilde{\infty}$ with neutral bands on either side.

When crystals of the mineral are heated much above 120° C they become cloudy, and the crystal structures seems at first sight to be lost, but by brightening up in oil it is found that a molecular rearrangement has taken place. This new mineral Rinne calls meta-scolecite. The inclined-faced hemihedrism of the monoclinic system is retained, but a remarkable revolution of the molecular groups through

[1]Edited by Dr. Wm. H. Hobbs, University of Wisconsin, Madison, Wis.
[2]Am. Jour. Sci., (3) xlvii, pp. 212–215.
[3]Neues Jahrb. f. Mineral., etc., 1894, II, pp. 51–68.

an angle of 90° about the 'c axis has taken place. The ortho-pinacoid has become the clino-pinacoid and vice-versa. The twinning
plane of twinned crystals has undergone the same revolution. By
heating crystals beyond the temperature required for producing the
first meta-scolecite, the double refraction of the substance steadily decreases and the symmetry approaches more and more closely to the
orthorhombic. Below red heat the structure breaks down. As scolecite
possesses three molecules of water of crystallization, Rinne suggests
that the first meta-scolecite contains two, the second one molecule of
crystal water, the crystal structure being lost when all the water has
been removed.

Crystallization of Herderite.—Penfield[4] has made a study of
herderite from the known localities as well as from a newly discovered
locality at Paris, Me.   The herderite from the latter locality as well
as that from Hebron, contains scarcely any fluorine, its place being
taken by hydroxyl, and the author proposes for it the name hydro-
herderite. As the Stoneham herderite contains hydroxyl and fluorine
in the proportions of 3 : 2, the one apparently replacing the other iso-
morphically, the name hydro-fluor-herderite is proposed for such intermediate varieties between theoretical fluor-herderite and hydro-herde-
rite. In the crystallographic study the fact is brought out that the
mineral is monoclinic instead of orthorhombic as has been supposed.
This is proven not alone on Paris specimens but on specimens from
the other localities, which were reexamined for this purpose. The
crystals, however, approach closely to the orthorhombic system, the
hydro- fluor-herderite being more nearly orthorhombic than the hydo-
herderite, the substitution of fluorine for hydroxyl tending to increase
the crystallographical axial angle and to shorten the clino-diagonal.
It likewise diminishes the mean index of refraction and the optical
angle.

Composition and Related Physical Properties of Topaz.—
Jannatsch and Locke[5] have shown that topaz contains water of constitution, from a chemical study of specimens from San Louis Potosi,
Ilmen Mts., Schneckenstein, and Brazil. Penfield and Minor[6] have
independently established the same fact by a larger number of analyses,
and shown how this greatly simplifies the formula of the mineral on

[4]Am. Jour. Sci., (3) xlvii, pp. 329–339.
[5]Am. Jour. Sci., (3) xlvii, pp. 386–387.
[6]Ibidem, pp. 387–396.

the assumption that hydroxyl and fluorine are isomorphous. Their results show that whereas the ratio $SiO_2 : Al_2O_3 : F$ varies from $1:1:1.50$ to $1:1:1.84$, the ratio $SiO_2 : Al_2O_3 : (F. OH)$ is constant and $1:1:2$, so the formula of topaz becomes $(Al [F. OH])_2 SiO_4$ or $(Al [F. OH]_2) Al SiO_4$. Their study of the physical properties of the mineral establishes a definite relation between them and the per cents of fluorine and water present, clearly indicating the isomorphous character of the fluorine and hydroxyl. The hydro-topaz has the smaller optical angle and the smaller specific gravity. The same fact is brought out by the determined values for a, β, and γ, and by exact measurements of interfacial angles. The optical anomalies of some Brazilian crystals are explained by zonal growth of topazes of different composition.

Composition of Chondrodite, Humite, and Clinohumite.— Penfield and Howe[1] have undertaken the study of the composition of the members of the humite group with the result not only of bringing order out of chaos, but also of establishing the fact that chondrodite, humite, and clinohumite constitute an homologous series both in a chemical and in a crystallographical sense. Sjögren has assumed that fluorine and hydroxyl are isomorphous, and derived new formulas for the members of this series, but as the authors point out the older analyses which Sjögren utilized are low as regards water, and Sjögren neglected to take into account the replacement of magnesia by ferrous iron and the consequent lowering of the silica percentage. The formulas derived by the authors, reckoning ferrous iron as magnesia, are as follows:

Chondrodite	$Mg_2 (Mg [F. OH])_2 (SiO_4)_2$
Humite	$Mg_4 (Mg [F. OH])_2 (SiO_4)_3$
Clinohumite	$Mg_7 (Mg [F. OH])_2 (SiO_4)_4$

The common difference of this homologous series is a molecule of chrysolite, $Mg_2 SiO_4$. As shown by Sacchi and vom Rath, if the 'c, axis of crystals of chondrodite be divided by 5, that of humite by 7, and that of clinohumite by 9, the axial ratios of the three minerals become practically identical. Now these divisors, 5, 7, and 9, are the same as the number of magnesia atoms in the formulas of the corresponding minerals. A most interesting relation is thus brought out connecting the crystal forms and chemical compositions of the members of this group. The authors think it probable that other members of this series will be discovered, such as a mineral of the composition $Mg (Mg [F. OH])$,

[1] Am. Jour. Sci., (3) xlvii, pp. 188–206.

SiO_4. This compound should have either orthorhombic or monoclinic symmetry, with β equal to 90° and an axial ratio a : b : c=1.086 : 1 : 1.887.

Leucite from New Jersey.—Kemp[8] argues for the presence of partially decomposed leucites in a dyke rock at Rudeville, Sussex Co., N. J., from a micro-chemical test indicating the presence of potassium, and from remains of leucite twinning, in spheroids now largely made up of analcite, calcite, feldspar, and other supposed secondary products.

Variscite from Utah.—Packard[9] gives an analysis of a specimen of compact or cryptocrystalline variscite from a quartz vein near Lewiston, Utah. The analysis is as follows:

H_2O 22.95 P_2O_5 44.40 Al_2O_3 (By difference) 32.65.

Utilization of Auerbach Calcite for Nicols.—An attempt has been made[10] to utilize the clear calcite from Auerbach on the Bergstrasse, Germany, for Nicol's prisms. Four ordinary Nicols with inclined end faces were prepared by Schmidt & Haensch of Berlin, and although these are equal to the medium quality Nicols prepared from Iceland spar in the matter of extinction, they nevertheless contain inclusions, air bubbles, etc., which are visible even to the naked eye. Dr. Hoffman, the owner of the Auerbach quarries, still hopes to secure material pure enough to take the place of Iceland spar. The material already tested will suffice for technical purposes.

Crystallization of Willemite.—Willemite has been supposed to have rhombohedral tetartohedral symmetry from the similarity of its rhombohedral angles to those of phenacite. Penfield[11] studies crystals from the Merritt Mine, N. M., Sedalia Mine, Salida, Col., and Franklin, N. J. In the specimens from the first and last mentioned localities, rhombohedrons of the second and third orders were observed and measured, showing that the system is what has been supposed. On the crystals from the Merritt Mine the second and third order rhombohedrons are $\frac{3}{4} P 2 \frac{1}{r}$ and $\frac{3}{4} P \frac{1}{4} \frac{r}{1}$ respectively. One of the types from the Franklin Mines is terminated by a third order rhombohedron $\frac{3 P \frac{3}{4}}{4} \frac{1}{r}$ alone, thus resembling the phenacite crystals from Mte. Antero, Col.

[8]Am. Jour. Sci., (3) xlvii, pp. 339–340.
[9]Am. Jour. Sci., (3) xlvii, pp. 297–298.
[10]Zeitschrift für Instrumentenkunde, 14te Jahrgang (1894), p. 54.
[11]Am. Jour. Sci., (3), xlvii, pp. 305–309.

The author shows that the cleavage of willemite is like that of troostite, indistinct cleavages parallel to both the base and prism being made out in willemite.

Composition of Staurolite and Arrangement of its Inclusions.—Exceptionally pure material for analysis was obtained by Penfield and Pratt[12] from St. Gothard, Switz., Windham, Me., Lisbon, N. H., and near Burnsville, N. C. A powder of uniform specific gravity was obtained in each case by the use of fused silver nitrate as a separating fluid in a specially constructed apparatus, the heavier and lighter portions of the powder being in this way removed. Reckoning MnO and MgO as FeO, and Fe_2O_3 as Al_2O_3, the four specimens yield results that agree well and indicate clearly that staurolite has the empirical formula H Al_5 Fe Si_2O_{13} as already suggested by Groth. The silica alone does not agree closely with this formula, being in every case about one per cent too high, and the authors think that this is due to the presence of inclusions of quartz too minute to be separated from the powder. Carbonaceous inclusions are in the staurolite from Lisbon, N. H., arranged in the same manner as in chiastolite crystals. The explanation of the authors is that the crystals of staurolite in growing in a solid rock, find it difficult to exclude foreign substances, the tendency to include them being greatest at the crystal edge and greatest where the interfacial angle is largest.

Determination of Quartz and the Feldspars in thin Section.—Sometime since Becke described a method of distinguishing quartz from feldspar by treatment with hydrochloric acid and subsequently tinting. He now[13] applies the same method to distinguish orthoclase from plagioclase and to determine the particular plagioclase species. Orthoclase is less affected by acid than plagioclase, and the soda rich plagioclases are less affected than the lime rich species. In rocks containing quartz, orthoclase and plagioclase, the slide is etched until by tinting the plagioclase shows an intense color. The orthoclase will then be faintly tinted and the quartz entirely unaffected.

Continuing his study Becke[14] has devised methods for the same determinations based on differences of refractive index. The first method consists in the examination of a perpendicular contact plane between

[12]Am. Jour. Sci., (3), xlvii, pp. 81-89.
[13]Tscherm. min. u. petrog. Mitth., xii, Heft 3, p. 2 (Notizen).
[14]Sitzungsber. d. k. Akad. d. Wissensch. i. Wien, Math. Naturw. Classe, Bd. 11, Abth. I, pp. 358-376, July, 1893.

the two minerals with a cone of illumination of small angle. When properly focused, this contact appears as a sharp line. On raising the tube of the instrument, the focus is disturbed and a light band appears on the side of the contact toward the more refractive mineral, which band widens and finally fades out as the tube is raised higher. If, on the other hand, the tube be lowered, the same phenomena appear on the other side of the contact. The best results are obtained with the use of high powers and with a cone of illumination of small angle. Becke recommends the use of the *Irisblende* furnished with the newer instruments of Fuess. I have obtained good results with a small Voigt and Hochgesang instrument by removing the weak convex lens which covers the polarizer. Becke's *Schlierenmethode* makes use of inclined illumination, which is obtained with the *Irisblende* or with Abbe's *Beleuchtungsapparat.* With inclined illumination, that side of a section of strongly refracting mineral toward the direction from which the light comes, shows a light band against the less strongly refracting mineral surrounding it, while the opposite side shows a dark band. The author states that this method suffices to determine orthoclase, quartz, and a plagioclase when they are present together in a holocrystalline rock, but suggests that it be supplemented by the *Färbung* method. The method of determining the species of plagioclase depends on the comparison of the double refraction of the feldspar with that of quartz sections. By making per cents of An the abscissæ, and indices of refraction the ordinates, curves are obtained for a, β and γ within the feldspar series. These curves are intersected by the horizontal curves of ω and ϵ in quartz. If now a' and γ' be the less and the greater values respectively of the refraction for the two principal directions in any section of plagioclase, a' being between a and β and γ' between β and γ, the curves obtained indicate the following relations:

	Parallel Position		Crossed Position		Composition.
I	$\omega > \epsilon'$	$\epsilon > \gamma'$	$\omega > \gamma'$	$\epsilon > \epsilon'$	Ab —Ab$_2$An$_1$
II	$\omega > \epsilon'$	$\epsilon > \gamma'$	$\omega = \gamma'$	$\epsilon > \epsilon'$	Ab$_3$An$_1$—Ab$_2$An$_1$
III	$\omega = \epsilon'$	$\epsilon > \gamma'$	$\omega < \gamma'$	$\epsilon > \epsilon'$	Ab$_3$An$_1$—Ab$_2$An$_1$
IV	$\omega < \epsilon'$	$\epsilon = \gamma'$	$\omega < \gamma'$	$\epsilon > \epsilon'$	Ab$_2$An—Ab$_2$An$_2$
V	$\omega < \epsilon'$	$\epsilon < \gamma'$	$\omega < \gamma'$	$\epsilon = \epsilon'$	Ab$_3$An$_3$—Ab An$_1$
VI	$\omega < \epsilon'$	$\epsilon < \gamma'$	$\omega < \gamma'$	$\epsilon < \epsilon'$	Ab$_1$An$_1$—An$_1$

It is seen that these subdivisions of the plagioclases correspond in a general way to the earlier one of Tschermak, I being albite, II and III oligoclase, IV and V andesine, while VI includes labradorite, bytounite and anorthite. As Tschermak's later and more equable subdivisien of the series has not been generally accepted, Becke thinks the harmony between his natural table and the older scheme of Tschermak a reason for retaining the original classification. The practical method of utilizing the results in his table, consists in finding contiguous sections of quartz and plagioclase which extinguish nearly parallel to one another. By means of the quartz wedge it is then determined whether the double refraction of these sections is of the same or of opposite sense. If the former, they are said to have parallel position and will indicate some of the relations of the first column of the table, and, if the latter, they have crossed position and their relations will correspond to something in the second column of the table. The quartz section always yields ω and a value varying but little from ϵ.

This method applies only to holocrystalline rocks which contain quartz, but it is a discovery of much importance which will doubtless be of much service in the study of the crystalline schists. The author has applied the method to the determination of the feldspar in many rocks of the Rosenbusch collection of B. Stürz, and printed his list of determinations. An excellent photogram also accompanies the paper.

Fluid Enclosures in Sicilian Gypsum.—The Cianciana gypsum contains cavities filled with liquid, some of which are 3 cm. in extent. Sjögren[15] has analyzed the liquid with the following results:

K_2O	Na_2O	CaO	MgO	Cl	SO_3	Total	O deducted for Cl_2	Cor. Total
2.1	40.9	4.1	3.9	44.9	14.1	110.0	10.1	99.9

Corresponding to

K_2SO_4	Na_2SO_4	$CaSO_4$	NaCl	$MgCl_2$	Total
3.7	11.4	9.7	66.2	9.0	100.0

The saline constituents were 4.023 per cent of the solution. This fluid is a fossil water of Miocene age, and differs from ocean water chiefly by containing a greater percentage of sulphates. It agrees fairly well with the water of some sulphur springs. The author thinks that the quantity of sulphates present in the water of the enclosure shows that the gypsum and sulphur cannot have been derived from a lagoon of sea water in which organic matters have reduced sulphur from the contained sulphates. Whether they are the product of sulphur springs or of emanations of H_2S in a lagoon of sea water in which sulphur has been deposited and sulphates formed by action of SO_3 on marls, the author is unable to determine.

[15]Bull. Geol. Inst. Upsala, I, (1898), No. 2, pp. 1-7.

New Sulphostannate from Bolivia.—In 1893 Penfield described a new isometric germanium mineral from Bolivia, which had the formula $Ag_8 Ge S_6$, and which he named canfieldite. This he showed to be identical chemically with Winkler's Freiberg mineral argyrodite, which that chemist had given the formula $Ag_6 Ge S_5$ and which Weisbach had considered monoclinic. Weisbach has since found that his earlier determination of the symmetry was incorrect, it being isometric tetrahedral and identical with the Bolivian mineral which should hence bear the name argyrodite. Penfield now transfers the name canfieldite[16] to a new sulphostannate of silver from La Paz, Bolivia, having isometric symmetry. A part of the tin is replaced by Germanium. The formula of the mineral is Ag_8 (Sn Ge) S_6, argyrodite being $Ag_8 Ge S_6$. The two minerals have similar physical properties, and are evidently isomorphous.

Allanite from Franklin Furnace.—Eakle[17] has made a crystallographical study of the allanite from the Trotter Mine, Franklin Furnace, N. J. The crystals occur in a granite dike associated with zinc ores. They are variable in habit and exhibit in all fourteen forms, none of which are, however, new to the species. The same author describes the tourmalines[18] from Rudeville and Franklin Furnace.

Miscellaneous.—Model[19] has found molybdenite and molybdite in the serpentine of the Rothenkopf, Zillerthal—. Carnot[20] has made an examination of the composition of wavellite and turquoise. In four analyses of wavellite from Cork, Ireland; Clomnel, Ireland; "Chester, Etats unis" (probably from Pennsylvania); and Garland, Arkansas, the fluorine was found to be 1.90, 2.79, 2.09 and 1.81 per cents respectively. Carnot proposes for the mineral the formula $2 (P_2O_5 Al_2O_3)+Al_2 (O_2F_6)+13 H_2O$, but in the light of the recent work of Penfield, it seems more probable that part at least of the water present, is water of constitution, and that the fluorine replaces hydroxyl and not oxygen. In two specimens of turquoise of mineral origin (from Persia and Nevada respectively) no fluorine was found. Two specimens of occidental turquoise (odontolite) yielded each over three per cent of fluorine. The entrance of fluorine into odontolite during its derivation from fossil teeth, the author was led to expect from his study of the composition of fossil bones of the different geological ages.

[16]Am. Jour. Sci., [3], xlvii, pp. 451–4.
[17]Trans. N. Y. Acad. Sci., xiii, p. 102; also Am. Jour. Sci., [3] xlvii, pp. 436–8.
[18]Am. Jour. Sci., [3], xlvii, p. 439.
[19]Tscheim. min. u. petrog. Mitth., xiii, p. 532.
[20]Comptes rendus. cxviii, pp. 995–8.

Reprinted from The American Naturalist, October 2nd, 1894.

General Notes.

PETROGRAPHY.[1]

Zirkel's Petrographie.—The second volume of Zirkel's treatise on Petrography[2] has recently appeared in America. It treats with such fulness of the massive rocks that an epitome of its contents is out of the question in this place. The volume discusses the composition, mineral and chemical, the structure and the distribution of the various types of the eruptive rocks with a thoroughness found only in German text-books. The descriptions of their important occurrences will be especially valuable to the student who has not a library at his disposal; and to the investigator, the large and accurate lists of references scattered through the book are very welcome. Many petrographers will differ with the author as to the importance and desirability of some of his types, and others will find fault with him concerning some of his theories, as, for instance, that of the origin of olivine aggregates in basalts. The volume is, however, on the whole quite free from theoretical discussions. While it loses something of its interest in consequence of this lack of theory, the book gains the confidence of the reader, who desires more particularly an account of the work done in the different provinces, where the rocks in which he is interested are to be found.

Inclusions in Volcanic Rocks.—Two articles on the petrographical changes affected by the partial or entire solution of foreign inclusions in volcanic rocks have recently appeared. The first is an essay by Dannenberg,[3] and the second a volume of 710 pages by Lacroix.[4] Dannenberg's article treats more particularly of the inclusions in the Siebengebirge basalts, andesites and trachytes. Zircons, corundum, magnetite, pyrite, feldspar, sillimanite, quartz, sandstone, schists and granite were found included in both basic and acid rocks of the region. Those inclusions that were most similar to the including rocks suffered much less alteration than those that differed most in chemical composition from the lavas. The aluminous compounds frequently yielded

[1] Edited by Dr. W. S. Bayley, Colby University, Waterville, Me.
[2] F. Zirkel: Lehrbuch der Petrographie, Leipzig, 1894, pp. iv and 941.
[3] Min. u. Petrog. Mitth. XIV, p. 17.
[4] Les Enclaves des Roches Volcaniques, Macon, 1893, pp. 710, pl. vii, fig. 84.

spinels as a consequence of the contact action. In many instances
different combinations of inclusion and including rocks gave rise to
the same new products, so that it is difficult to discover the exact law
governing the changes. In the basalts the principal inclusions consisted of single minerals, while in the more acid rocks they comprised
largely rock fragments—a fact probably attributable to the different
solvent powers of the including material. Lacroix's volume is a
nearly complete treatise on the subject of which it treats, which is limited, as the title indicates, to the study of inclusions in volcanic (effusive) rocks only. The author separates inclusions into two classes.
The first comprises fragments of an entirely different nature from that
of the enclosing rock, as granite in basalt. These he calls enallogenous (enclaves énallogènes). The second class comprehends inclusions
more or less similar in composition to the including material. These
he terms homogeneous inclusions (enclaves homoeogèues). The second
class embraces aggregates formed by segregation and by liquation, as
well as true inclusions. The including rocks are also separated into
two groups, the basaltic and the trachytic. In the first part of the
book the enallogenous inclusions are discussed with great thoroughness.
In the second part the homogenous inclusions are studied. In a third
part are collected the general conclusions. Chapters are devoted to
each class of rocks and divisions of the chapters to the character of
the inclusions in them. Resumés and paragraphs embracing the
results of the studies are scattered through the volume at convenient
intervals, and a geographical index concludes the book. The number
of discoveries made by the author in the course of his work is too
large for discussion in this place. The book bears evidence of thoroughness throughout. It is an excellent contribution to the subject of contact action.

**The Basic Rocks of the Adirondacks and of the Lake
Champlain Region.**—Kemp[5] gives a brief account of the coarse
basic rocks of the Adirondacks of which the well known norite is a phase
Associated with the norite are anorthosites, gabbros and olivine gabbros,
all of which are more or less schistose. The anorthosites are crushed,
and where the shattering has been most intense their plagioclase has been
changed to a fine grained aggregate, thought to be saussurite. Augite
and brown hornblende are present in these rocks, but not in large
quantity. Garnets are always present. The more basic gabbros are
dark rocks, whose plagioclase has a greenish tinge due to the abund-

[5] Bull. Geol. Soc. Amer. 5, p. 213.

ance of dust inclusions scattered through it. The special features of the gabbros are the reaction rims around pyroxene and magnetite. A zone of small brown hornblendes is often found between the first named mineral and plagioclase. Between magnetite and feldspar are usually three zones, of brown hornblende, pink garnet, and quartz, respectively, the last named mineral occurring nearest the feldspar. Sometimes the order of the zones is different. The quartz may appear within the zone of garnets, in which case the latter mineral may replace the feldspar in part, as alternate lamellae between lamellae of plagioclase. The gabbros contain large bodies of titaniferous magnetite. On the contact of the eruptive with limestone the latter rock has been crystallized and silicified. The same author, associated with Marsters,[6] has described the trap dykes of the Lake Champlain region as camptonites, fourchites, monchiquites and bostonites.

The Augite Granite of Kekaquabic Lake, Minnesota.— The granite of Kekaquabic Lake in Northeastern Minnesota, occurs in granitic and in porphyritic phases, according to Grant.[7] In both varieties the constituents are quartz, anorthoclase and other feldspars, augite, a little hornblende, biotite, apatite and sphene. The granitic variety needs no further mention. In the porphyritic phase the quartz and feldspar form a fine grained groundmass in which lie phenocrysts of feldspar and augite. An analysis of this feldspar, whose density is 2.58–2.62 gave:

SiO_2	Al_2O_3	Fe_2O_3	CaO	MgO	K_2O	Na_2O	H_2O	Total
67.99	19.27	.82	.75	.02	3.05	6.23	.90	= 99.03

The augite comprises from 5–20 per cent of the rock. Its tint varies from green to colorless, the lighter colored portion often lying within a darker outer zone. Analysis of the augite yielded:

SiO_2	Al_2O_3	Fe_2O	FeO	CaO	MgO	K_2O	Na_2O	H_2O	Total
53.19	2.38	9.25	5.15	17.81	9.43	.38	2.63	.01	= 100.23

The rock, which is an augite soda-granite, has the following composition:

SiO_2	Al_2O_3	Fe_2O_3	FeO	CaO	MgO	K_2O	Na_2O	H_2O	P_2O_5	Total
66.84	18.22	2.27	.20	3.31	.81	2.80	5.14	.46	tr	= 100.05

[6]Bull U S. Geol. Survey, No. 107.
[7]Amer. Geol., XI, 1893, p. 383.

Petrographical News.—In a series of articles recently published Vogt[8] discusses the formation of oxides and sulphide ores around basic eruptive rock bodies, describes all the known occurrences of the nickel sulphides with reference to their mode of origin, and reviews critically the literature treating of the differentiation of rock magmas. He shows that the nickel ore deposits that are peripheral must be due to differentiation of rock magmas. He further shows that the laws governing the processes of differentiation are very complicated and that neither Soret's principle nor any other single physical or chemical principle will satisfactorily explain the phenomena.

Dr. G. H. Williams[9] reports the occurrence of volcanic rocks at many localities in the eastern crystalline belt of North America. The rocks in question comprise tuffs, glass breccias, devitrified obsidians and fine grained crystalline flow rocks with many of the characteristics of modern lavas. All these have heretofore been regarded as sedimentary in origin by most of the geologists who have studied them. The author gives his reasons for concluding that they are volcanic, and declares that, not before their true character is recognized will the structure of the crystalline areas of the Appalachians be correctly understood.

Lang[10] discusses the conclusions of Rosenbuch[11] with respect to the chemical nature of the crystalline schists, and criticizes Linck's principles governing the mineralogical composition of eruptive rocks. In his article, which is well worth reading, the author shows conclusively that the mineral composition of rocks is not determined by their chemical composition.

[8]Zeits f. prakt. Geol., 1893, Jan., April.
[9]Journ. of Geol., Vol. 2, p. 1.
[10]Min. u. Petrog. Mitth., XIII, p. 496.
[11]Cf. American Naturalist, 1891, p. 827.

Reprinted from The American Naturalist, November 1st, 1894.

General Notes.

PETROGRAPHY.[1]

Composite Dykes on Arran.—Professor Judd[2] describes a number of "composite" dykes on the Island of Arran, in which the well-known "Arran pitchstone" and a glossy augite-andesite occupy different portions of the same fissure, either rock appearing in the center of the dyke, with the other on one or both of its peripheries, or the one rock cutting irregularly through the other. The relations of the rocks indicate that there was no regular sequence in the intrusion, the pitchstone having been intruded sometimes before, sometimes after the andesite. Each rock contains fragments of the other (in different dykes), and the two rocks are always separated by a sharp line of demarkation. The andesite is a basic rock containing about 56 per cent of silica, while the pitchstone is a pantellerite with 75 per cent of SiO_2 or an augite-enstatite-andesite with 66 per cent of SiO_2 and 4.13 per cent K_2O. The andesite is well characterised. It passes into a tholeite with intersertal structure, by a decrease in the glassy component, and upon further loss of glass it passes into diabase. The pitchstone is largely an acid glass, surrounding crystals of quartz, and microlites of augite, feldspar, magnetite, etc. The author adds to the list of individualized components already known to exist in the rock hyalite and tridymite. The latter mineral occurs in plates aggregated into spherules and globules that surround quartz crystals, and the hyalite forms globules scattered here and there through the glass. The author thinks that materials of such widely different nature as that existing in these dykes could not have been formed by the differentiation of a magma after its intrusion into the dyke fissures, but that the differentiation must have taken place while the magma was still in its subterranean reservoir.

Analyses of Clays.—Hutchings[3] quotes a series of analyses of carboniferous clays to show that these substances possess the requisite composition to become clay slates upon compression. He ascribes the small percentages of alkalies shown in most clay analyses to the fact

[1] Edited by Dr. W. S. Bayley, Colby University, Waterville, Maine.
[2] Quart. Jour. Geol. Soc., xlix, 1893, p. 536.
[3] Geol. Magazine, Jan. and Feb., 1894.

that these analyses are of commercially valuable clays, selected for their small alkali contents. In the course of his article the author corrects some of the statements made in earlier papers and amplifies others. He declares that newly formed feldspar is present in the slates metamorphosed[4] by the shap granite and in other contact slates. In the spots of the shap rocks, and in those of other contact slates, there is always present, in addition to its individual components, more or less of a yellowish-green very weakly polarizing substance in which the other components of the spot are imbedded. This is believed to possess an indefinite composition, and to be the result of aqueo-fusion of some of the constituents of the original rock and the solidification of the product in an amorphous condition. The paper concludes with a statement of the author's views concerning the transformations that rutile, biotite, quartz, feldspar, cordierite and other contact minerals undergo in cases of contact metamorphism.

The Phonolites of Northern Bohemia.—The phonolites of the Friedländer district of North Bohemia are nosean bearing trachytic phonolites and nepheline-phonolites, according to Blumrich.[5] The latter contain phenocrysts of anorthoclase in a groundmass of sanidine, nepheline and aegerine crystals and groups of a new mineral which the author calls hainite. This hainite is a strongly refracting but a weakly doubly refracting colorless substance. It occurs in tiny triclinic needles with a density of 3.184. These unite into groups. It is found also as well-developed wine-yellow crystals forming druses in cavities in the rock. The mineral has a hardness of 5, and it is optically positive. It is supposed to be closely related to rinkite, hjortdahlite and the other fluorine bearing silicates common to the eleolite-syenites. In addition to hainite the druse cavities contain albite, chabazite and nosean. In the trachytic phonolites a glassy base was detected.

Spherulitic Granite in Sweden.—Loose blocks of spherical granite are reported by Backström[6] from Kortfors, in Orebro, and Balungstrand in Dalekarlien, Sweden. The rock from Kortfors is a hornblende granitite containing concentric nodules composed of four zones. The inner one consists of oligoclase, microcline and quartz; the second of oligoclase in radial masses and small quantities of hornblende, biotite, magnetite, orthoclase and quartz; the third of hornblende, biotite, oligoclase and a little biotite, and the peripheral zone

[4] Cf. American Naturalist, 1892, p. 245.
[5] Min. u. Petrog. Mitth., xiii, p. 465.
[6] Geol. Foren. i. Stockh. Förh. 16, p. 107.

of magnetite in a matrix of oligoclase. The structure of the spheroids, with the younger minerals nucleally and the older ones peripherally distributed, indicates to the author that they were produced by liquation processes. The rock from Balungstrand possesses a coarse ground-mass consisting almost exclusively of microcline and quartz. The spheroids are essentially oligoclase spherulites peripherally enriched by biotite. They are clearly older than the groundmass.

Diabase and Bostonite from New York.—A few dyke rocks cutting the gneisses of Lynn Mountain, near Chateaugay Lake, Clinton Co., N. Y., are described by Eakle[7] as consisting of olivine diabase and of bostonite. The latter rock is porphyritic with phenocrysts of red orthoclase in a fine-grained groundmass with the trachytic structure. It differs from the other bostonites of the region in the presence of much chloritized augite in its groundmass. It is also more acid than these. Its analysis gave:

SiO$_2$	Al$_2$O$_3$	Fe$_2$O$_3$	CaO	MgO	K$_2$O	Na$_2$O	Loss	Total
67.16	14.53	4.17	1.26	.41	6.10	5.55	1.10	= 100.28

The olivine diabase differs from the ordinary ophitic diabases in that much of its augite is in idiomorphic forms. They thus resemble Kemp's augite camptonites.

Petrographical News.—A very interesting series of analyses of rocks from the central and northeastern portions of the Mittelgebirge is given by Hibsch.[8] The series includes analyses of phonolites, dolerites, camptonites, nepheline and leucite tephrites, augitites and basanites. Many of the rocks have been described in the literature.

Cohen[9] has obtained from the Transvaal, Africa, specimens of a calcite bearing aplite and of a melilite augite rock of a somewhat abnormal character. The aplite is from the mine of the Iron Crown Gold Mining Co., near Hamertsburg, and the melilite rock from near Palabora. The melilite rock is a fine-grained aggregate composed largely of honey-yellow melilites and black augites. On its druse walls are little crystals of the first-named mineral, and through the druse cavities extend thin plates of copper. In the thin section, clear, colorless melilites, with rounded outlines and olive-green grains of augite are seen to lie in an opaque granular groundmass in which are dots and flakes of copper.

[7] Amer. Geologist, xii, p. 31.
[8] Min. u. Petrog. Mitth., xiv, p. 95.
[9] Minn. u. Petrog. Mitth., xiv, p. 188.

Backström[10] fused feldspathic phonolite and obtained as the product upon cooling a colorless glass filled with microlites of oligïoclase, nepheline, small microlites of colorless pyroxene and tiny grains of picotite and olivine (?). Upon fusing a leucite phonolite, containing nosean, SO_3 is driven off and the resulting product is a glass enclosing microlites of oligoclase, a few prisms of nepheline and abundant crystals of a yellow pyroxene with the properties of aegerine.

[10] Bull. d. l. Soc. Franc. d. Min., 1893, xvi, p. 130.

Reprinted from The American Naturalist, December 1st, 1894.

PARTIAL INDEX OF SUBJECTS.

A
SUMMARY OF PROGRESS

IN

PETROGRAPHY

IN

1895.

BY

W. S. BAYLEY.

FROM MONTHLY NOTES IN THE "AMERICAN NATURALIST."

PRICE 50 CENTS.

WATERVILLE, ME.:
PRINTED AT THE MAIL OFFICE,
1896.

A

SUMMARY OF PROGRESS

IN

PETROGRAPHY

IN

1895.

BY

W. S. BAYLEY.

FROM MONTHLY NOTES IN THE "AMERICAN NATURALIST."

PRICE 50 CENTS.

WATERVILLE, ME.:
PRINTED AT THE MAIL OFFICE.
1896.

PETROGRAPHY.[1]

The Serpentines of San Francisco.—The serpentine of the Protero, a district within the limits of the city of San Francisco, is an eruptive rock intrusive in sandstone. It was originally a lherzolite, which by the usual processes of alteration has been changed to serpentine. Two varieties of the rock are noticed by Palache.[2] One is a massive form, while the other is slickensided along so many planes close together that the rock has became schistose. Between the slickensided surfaces are often spheroidal masses of the massive rock. The massive serpentine is of the usual character. It consists now of a felt of serpentine fibres in which are imbedded numerous crystal-like areas of enstatite and diallage, and grains of olivine, magnetite and chromite. The crystal-like particles of the pyroxenes are remnants of larger grains that were shattered by dynamic action. The pyroxenes and the olivine have yielded the serpentine. Intrusive into the serpentine is a hypersthene diabase, composed of labradorite, monoclinic and orthorhombic pyroxenes and green hornblende, supposed to be derived from the pyroxene. Its structure is ophitic. A second variety of the rock consists essentially of plagioclase and hornblende. Portions of it are schistose. Its structure is sometimes granitic and sometimes ophitic, and in the latter case it contains small quantities of pyroxene. Hence it is regarded as an altered form of the diabase. An analysis of the hornblende variety follows:

SiO_2 Al_2O_3 Fe_2O_3 FeO MnO CaO MgO K_2O Na_2O P_2O_5 TiO_2 H_2O Total
47.41 16.03 2.66 7.05 tr 12.33 5.81 4.47 tr 1.29 2.19=99.24
Density = 2.96.

The Blue Hornblende in the California Schists.—In many of the schists of the Coast Range, Cal., is a blue amphibole that has for some years past gone under the name of glaucophane. Palache[3] has recently found it in large quantities and in well developed columnar crystals in a schist-boulder near Berkely. The matrix of the schist is a granular aggregate of clear, fresh albite, containing numerous liquid and solid inclusions. The latter consist largely of small grains

[1] Edited by Dr. W. S. Bayley, Colby University, Waterville, Maine.
[2] Bull. Geol. Dept. Univ. of Cal., Vol. 1, p. 161.
[3] Ib., Vol. 1, p. 181.

and needles of the blue amphibole. In addition to these are tiny crystals of magnetite, sphene and zircon. In this matrix lie sheaves of the blue amphibole, which are formed of small needles or of large columnar crystals, sometimes measuring as much as 20 mm. in length. The crystals are well developed in the prismatic zone, where they exhibit clearly the cross section of amphibole. The plane of their optical axes is the clinopinacoid. The extinction of the mineral is about 13° to c, along the axis of greatest elasticity. The mineral must be closely related to riebeckite. A characteristic feature of the new amphibole is its strong pleochoism, which is stronger even than that of glaucophane. A=sky blue to dark blue; B=reddish to purplish-violet; C= yellowish-brown to greenish-yellow. When broken, crystals of the blue amphibole are often healed with green actinolite, and often fibres of the latter mineral unite portions of blue crystals on opposite sides of veins of albite cutting through the rock mass. An analysis of the blue mineral gave:

SiO_2	Al_2O_3	Fe_2O_3	FeO	MnO	MgO	CaO	Na_2O	K_2O	H_2O	Total
55.02	4.75	10.91	9.46	tr	9.30	2.38	7.62	.27	undet.	= 99.70

This indicates a mixture of the three molecules $Na_2 Al_2 Si_4O_{12}$, $Na_2 Fe_2''' Si_4O_{12}$ and $R''SiO_3$ (where R is Mg : Fe : Ca $= 6:2:1$) in the proportions $1:2:9$. The optical properties of the mineral are very similar to those of the blue amphibole described by Cross.[4] Chemically, it lies between riebeckite and glaucophane. The author names it crossite.

The Diorites, Gabbros and Amphibolites of Argentina.—

The basic rocks from Argentine in the collection of Berlin University have been studied petrographically by Romberg.[5] They occur in the easternmost of the Cordilleran chains, associated with crystalline schists and eruptive rocks. The diorites and gabbros form stocks, and sometimes sills and dykes, that are closely associated with gneiss and crystalline limestones. The author divides the rocks studied into a number of groups and sub-groups, recognizing as the two principal groups eruptive rocks, and those associated with the crystalline schists. Among the undoubted eruptives are gabbros and diorites, and of the latter class there are two varieties, the diorites proper and the quartz diorites. Gabbro-diorites are also recognized among the specimens. The gab-

---

[4] Cf. American Naturalist, 1890, p. 1073.
[5] Neues Jarb. f. Min. etc., B. B., ix, p. 293.

bros include olivinitic and non-olivinitic varieties. In the former there is often a bluish green hornblende, at whose contact with feldspar there is often a fringe of spinel arranged in pseudopodia-like masses with their long directions perpendicular to the bounding surfaces of the amphibole. In other specimens the olivine is separated from feldspar by a band of hypersthene. Norites, with reaction-rims around their olivines, and peridotites containing enstatite are among the other members of the gabbro family met with. More closely associated with the schists than all the rocks just mentioned, and apparently forming a portion of the schist series, are diorites, often saussuritized, and amphibolites among the hornblende rocks, and gabbros, peridotites and serpentines among the pyroxene bearing kinds. The basic schistose rocks in the collection studied are schistose diorites, and rocks composed essentially of epidote and zoisite, and of garnet and scapolite, supposed to be derived from diorite, schistose gabbros and hornblende schists. After describing the characteristic features of the gabbro and diorite structures, the author proceeds to discuss the origin of the Argentine hornblende schists. He finds no evidence that these are squeezed plutonic rocks nor metamorphosed sediments, and so he concludes that they are submarine eruptives.

Amphiboles in Russian Rocks.—Federow[*] gives some interesting notes on the amphiboles in the rocks of the northern Urals. The mineral is frequently absent from the freshest rocks. It is most abundantly present in those that have been metamorphosed by pressure. The kinds observed were a yellow-green variety, a colorless or very light colored kind, a dark brown variety, a fibrous variety with a blue color, glaucophane and gastaldite. The first is especially common in gneiss, syenite and syenitic gneiss, and it is present also in a diabase, where it is believed to have been derived from chlorite. The second variety is common to highly metamorphic rocks, while the third is limited to diabases and proterobases. The fourth variety is characteristic of the green schists, more particularly those that have undergone chemical alterations. The glaucophane is found in magnetite schists, in a few altered green schists and in gneiss. The sixth variety is also common to the green schists. In a syenite gneiss the author observed a brown augite that along a zone of crushing has been changed to a light green pyroxene, which is regarded as evidence that dark brown amphibole may give rise by pressure to light green hornblende.

[*] Minn. u. Petrog. Mitth., xiv, p. 143.

Basalt Boulders from Thetford, Vt.—A brief description of the material of the peculiar basalt boulders discovered by Hubbard at Thetford, Vt., is given by Hovey[7] in a recent paper. The most conspicuous features of the boulders are the large masses of olivine and pyroxene scattered through them. The former are in rounded aggregates with a granular structure. Their composition is $SiO_2 = 40.75$, $FeO = 9.36$; $MgO = 50.28$. The pyroxene nodules consist of the remnants of single crystals of a pale green color, and with an extinction of 44°. These nodules are in a groundmass composed of augite, plagioclase, hornblende and several accessory substances. The augite of the groundmass is brownish-violet in color, and it has the peculiarities of basaltic augite.

Maryland Granites.—Keyes[8] argues the original character of much of the epidote in Maryland granites from its close association with allanite, which is believed to be an original component of the rocks, since it occurs in them as sharply defined crystals completely mantled by fresh biotite. It is found also included in crystals of sphene of whose primary nature there can be no doubt. Finally its grains are idiomorphic with respect to many of the original rock components with which they are in contact.

[7] Trans. N, Y. Acad. Sciences, xiii, p. 161.
[8] Bull. Geol. Soc. Amer., Vol. 4, p. 305.

Reprinted from The American Naturalist, January 1st, 1894.

PETROGRAPHY.[1]

Geology of Angel Island, San Francisco Bay.—Angel Island in San Francisco Bay, Cal., consists essentially of a syncline of sandstone interbedded with an intrusive sheet of fourchite and cut by a serpentine dyke and a second mass of fourchite. A radiolarian chert is associated with the sandstone. The most interesting feature connected with the rocks is the discovery by Ransome[2] that both the fourchite and the serpentine have effected metamorphic changes in the sandstone and in the chert, and that in all cases the resulting product is the same, viz., a glaucophane schist. The serpentine and the fourchite are thus true eruptive rocks, neither being, as supposed by Becker, a metamorphosed sediment. The glaucophane schists are true contact rocks, and are not the result of a general oregional metamorphism of pre-existing rocks. Not only do they occur as contact facies of the sandstones and cherts, but the former rock often contains pebbles of schists, in their essential features similar to the contact schists. The sandstone is made up of quartz, plagioclase and fragments of various rocks. The fourchite consists almost entirely of nearly colorless augite in rounded or irregular grains, and a small quantity of an interstitial substance composed of smaller granules of augite and a fine grained neatrix, which under high powers resolves itself into small, stout colorless crystals imbedded in a yellowish-green substance that is nearly isotropic. The crystals are thought to be zoisite, which may be an alteration product of plagioclase, although the author thinks this origin not probable. Often the augite is changed peripherally into glaucophane, which either replaces the pyroxenes, fills cracks in them, or occurs in the spaces between adjacent grains. A few of the specimens examined possess a glassy groundmass and others are porphyritic. Brecciated and spheroidal facies were also observed. The schist produced by the alteration of the sandstones and cherts is sometimes composed of aggregates of glaucophane in a matrix of colorless albite. Brown mica, garnets and sphene are also present to some extent in the rock. Other varieties of the schist are essentially aggregates of quartz and glaucophane. Occasionally the glaucophane is in fairly well defined crystals, but usually it is in sheaf-like bundles of fine needles. The altered cherts now consist of spherules of cryptocrystalline silica

[1] Edited by Dr. W. S. Bayley, Colby University, Waterville, Maine.

[2] Bull. Dept. Geol. Univ. of Cal., Vol. 1, p. 193.

or grains of recrystallized quartz and microlites and bundles of nearly colorless augite. The serpentine on the island is nodular as the result of shearing. It was derived in all probability from a rock made up almost exclusively of diallage. It contains granules of chromite. This serpentine has effected the same alterations in the chert and sandstone through which it cuts, as has the fourchite. Some peculiar inclusions of a dark rock in the serpentine are supposed to be the remnants of a dyke that formerly occupied the fissure, which the serpentine subsequently filled. These fragments now consist of a holocrystalline aggregate of augite and albite, of which the first mineral is sometimes altered to green and brown hornblende. Analyses of the fourchite (I) and of a fresh nodule of serpentine (II) follow:

	SiO_2	Al_2O_3	Fe_2O_3	FeO	CaO	MgO	K_2O	Na_2O	P_2O_5	Loss	Total
I.	46.98	17.07	1.85	7.02	12.15	8.29	.53	2.54	.09	4.86	=101.38
II.	42.06		2.72		2.88		39.53	not estimated		12.04	= 99.23

A New Rock-Volcanite.—In an abstract of a paper to appear in a German periodical, Hobbs[3] gives an account of an anorthoclase-augite rock which he calls volcanite. It occurs as bombs projected from Volcano in 1888–89. Phenocrysts of anorthoclase, andesine, an acmitic augite and olivine are imbedded in a groundmass containing two generations of the first named of these minerals in a glassy base. The augite phenocrysts in many instances have been resorbed by the rock's magma and have thus given rise to pseudomorphs of colorless pyroxene, magnetite, and plagioclase. The structure of the rock is trachytic. Its chemical composition corresponds with that of the dacites, while its mineralogical composition is that of an augite pantellerite. Analyses of the anorthoclase (II) and of the rock mass (I) follow:

	SiO_2	Al_2O_3	Fe_2O_3	FeO	MnO	CaO	MgO	Na_2O	K_2O	H_2O	P_2O_5	Total
I.	66.99	17.56	1.41	3.39	tr	4.25	.93	3.35	.34	1.53	tr	= 99.75
II.	60.01	20.12		2.82		5.15	.23	6.43	3.67	.77		= 99.20

Acmite Trachytes from Montana.—Among the eruptive rocks occurring as dykes, sheets and laccolitic masses in the Cretaceous of the Crazy Mts. are acmite trachytes and eleolite syenites. The former, according to Wolff,[4] is present in small sheets and dykes and in

[3] Bull. Geol. Soc. Amer., Vol. 5, p. 598.
[4] Bull. Mus. Comp. Zool., xvi, p. 227.

apophyses from laccolitic masses. It is a rock made up of phenocrysts of anorthoclase, sodalite and augite in a groundmass of lath-shaped feldspars and acicular aegirines and acmites imbedded in a colorless interstitial matter, composed in all probability of nepheline and analcite. The augite phenocrysts are provided with an outer zone of aegerine. Needles of this mineral are included in all the colorless constituents. The eleolite syenite is from the laccolites. It is panidiomorphic, with fresh onorthoclase phenocrysts in a fine grained mass of feldspar, augite, aegerine, acmite, the angular spaces between which are occupied by nepheline. Analyses of the syenite and of one variety of the trachyte gave:

SiO_2	Al_2O_3	Fe_2O_3	FeO	MnO	CaO	MgO	Na_2O	K_2O	TiO_2	P_2O_5	Ign.	Loss	Total
59.66	16.97	3.18	1.15	.19	2.32	.80	8.38	4.17	tr	.14	2.53	.07	99.56
62.17	18.58	2.15	1.05	tr	1.57	.73	7.56	3.88	tr	.11	1.63	.07	99.50

Petrographical Notes.—In a glassy rock from near Harrismith, in the Orange Free States, Molengraff[5] finds small crystals of twinned cordierite, little octahedra of magnetite and skeleton crystals of augite. The cordierite is slightly pleochroic. Its crystals are well defined and possess all the peculiarities of the mineral. An analysis of the rock shows:

SiO_2	TiO_2	Al_2O_3	Fe_2O_3	FeO	MgO	CaO	K_2O	Na_2O	Loss	Total
64.54	.79	19.16		7.23	3.39	2.47	.57	1.13	2.25	=101.53

The large percentage of SiO_2 present as compared with the small percentages of the alkalies suggests to the author that the rock is an abnormal type. After a critical discussion of the literature of cordierite as a rock component, the conclusion is reached that, in all probability, the specimens studied represent foreign inclusions fused in a basic rock.

A very brief account of the lavas and ashes of the old volcano Rhobell Fawr near Dolgelley in Wales, is given us by Cole.[6] The greater portion of the products are ashes containing hornblende and augite. The lavas are augite, aphanites and basaltic and andesitic andesites.

[5] Neues Jarb. f. Min. etc., 1894, I, p. 79.

[6] Geol. Magazine, x, 1893, p. 337.

Rutley[1] gives a few illustrations in proof of his statement that the production of spherulites is sometimes a devitrification process subsequent in point of time to the development of perlitic cracks in the volcanic rocks in which they occur.

In a recent number of Science Blake[8] suggests the notion that many of the quartz veins, 'reefs' and boss-like masses in ancient rocks are the result of deposition from old thermal springs.

The rock of Saint Sardoux, Puy-de-Dame, France, is composed[9] of ilmenite and soda-augite in a groundmass consisting largely of nepheline crystals cemented by a matrix of feldspar and glass. Sometimes the augite and nepheline are intergrown like the constituents of a pegmatite and at other times they form an ophitic aggregate, with the nepheline the older component. The rock penetrates the peperites of the region in the form of dykes and veins.

New Books.—Granites and Greenstones[10] is the title of a new series of tables for the determination of rocks and their essential components. The author, Mr. Rutley, divides rocks into Volcanic rocks, Dykes and Sills and Plutonic masses, and then subdivides each group into four series as follows: Ultra-basic with $SiO_2 = 39–45$ per cent; basic, with silica 45–55 per cent, intermediate, with $SiO_2 = 55–66$ per cent; and acid, with silica over 66 per cent. The ultra-basic series is divided into the non-feldspathic and the potentially feldspathic, including nepheline and leucite non-feldspathic rocks. The basic rocks are all plagioclastic. They include a nephelinic or leucitic and a non-nephelinic group. In the dyke and sill division of this series are included the diabases. In the intermediate series we find again two groups—the orthoclastic and the plagioclastic, and in each of these nepheline and non-nepheline sub-groups. The acid series includes a division whose feldspars are plagioclase or anorthoclase, and one in which the feldspar is orthoclase. Definitions and notes are abundant and are so given as to really explain the tables. The elvans are described as the apophyses of deep seated granitic masses. They include micro-granites, aplites, quartz-porphyries and greisens. The mineralogy tables contain no startling novelties. They are good and the book itself is well worth study. It will serve as a useful companion to the student.

[1] Quart. Jour. Geol. Soc., Feb. 1894, p. 10.

[8] Science, Vol. xxiii, 1894, p. 141.

[9] Bull. Soc. Franc. d. Min. xvii, p. 43.

[10] Granites and Greenstones, a series of Tables and Notes for Students of Petrology. By Frank Rutley. London, Thos. Murby, 1894, pp. 48.

In an article of 52 pages on the optical recognition and economic importance of the common minerals found in building stones, Luquer[11] mentions the principal microscopic characteristics of the most important minerals with sufficient fullness to enable the technical student to determine them in the thin section. He also notes the effect of presence of each upon the value of the various building stones in which they occur. He, moreover, describes the usual associates of each different mineral, and so incidentally gives the composition of the principal rocks used in constructions. The article in pamphlet form is simple and useful.

[11] School of Mines Quart. xv, p. 285–336.

Reprinted from The American Naturalist, February 1st, 1895.

PETROGRAPHY.[1]

Some Basalts of Asia Minor.—The rocks near Kula, Asia Minor, are basalts in sheets and lava streams, the latter emanating from a number of old volcanic centers whose cores may still be distinguished. These basalts, according to Washington,[2] are hornblende-plagioclase basalts, characterized especially by the abundance of their hornblendic component. This mineral, augite and olivine are present as phenocrysts in a groundmass made up of plagioclase, magnetite and glass, the latter being lighter in color as the magnetite in it increases in quantity, thus indicating that this mineral was one of the latest separations from the magma. Leucite was discovered in two of the streams. It presents no unusual features. The mineral is rare in hornblendic basalts elsewhere. None of the components of the rocks merit special mention but the hornblende. This is always porphyritic and is present in large quantity. Its color is yellow, brown or greenish-yellow, and its extinction varies from 4° to 23°. The chemical alterations effected in the mineral by magmatic resorption are interesting. One effect is the replacement of the hornblende by a reddish-brown mineral associated with colorless augite and opacite, and another is its partial or complete alteration into augite and opacite. The brown mineral is referred to hypersthene, although the analysis of a portion of the rock containing a large quantity of it was rather against this theory. The author thinks that the formation of the mineral was probably due to the reducing action of hydrogen (from dissociated water included in the lava) upon the ferric iron of the hornblende. In structure the basalts are normal, hyalopilitic, semi-vitreous and tachylitic. An analysis of a leucite variety gave:

SiO_2	Al_2O_3	Fe_2O_3	FeO	CaO	MgO	Na_2O	K_2O	P_2O_5	H_2O	Total
47.74	20.95	3.29	6.32	7.56	5.16	7.12	1.21	.13	.04	= 99.52

Since the hornblende is of primary importance in the basalts of Kula, it is proposed to call them, and other basalts in which hornblende predominates over augite and olivine, by the name of Kulaites.

The Igneous Rocks of the Eureka District.—In an appendix to the Geology of the Eureka District, Iddings[3] gives an account of

[1] Edited by Dr. W. S. Bayley, Colby University, Waterville, Maine.
[2] Amer. Jour. Sci., Feb., 1894, p. 114.
[3] Monograph XX U. S. Geol. Survey, p. 337.

the igneous rocks of the region with special reference to the lavas
whose studies led Hague to the proposal of the theory that the various
types of rocks in the Eureka district are differentiated portions of one
magma, which split up into two, one yielding feldspathic acid rocks
and the other pyroxene basic ones. Among the intrusive rocks of the
region Iddings mentions only granites, granite-porphyries and quartz-
porphyries. The volcanic rocks include hornblende-andesite, horn-
blende-mica-andesite, dacite and rhyolite, which are the types derived
from the more acid portion of the original magma, and pyroxene-
andesites and basalts derived from the basic portion. The pyroxeneandesite contains anorthite, hypersthene, augite, hornblende, a little
biotite and an occasional quartz grain, in a glassy groundmass with a
felt-like structure produced by labradorite and augite-microlites. The
hornblende-mica-andesites are more acid. They contain labradorite,
hornblende, biotite and a little quartz as porphyritic crystals in a
micro-crystalline groundmass of lath-shaped plagioclases and intergrowths of feldspar and quartz. The dacites are rare. They possess
macroscopic quartz-phenocrysts together with hornblende, hypersthene,
a little augite, biotite, labradorite, anorthite, and possibly orthoclase
in a pumiceous glass base, which also often contains many beautifully
crystallized zircons. The rhyolites met with present few characters of
special interest. They vary in the texture of their groundmass from
micro-crystalline to glassy varieties. Their sanidine phenocrysts have
the plane of their optical axes sometimes in the plane of symmetry
and sometimes perpendicular thereto. Occasionally the rock possesses
also phenocrysts of hypersthene. The basalts are poor in olivine, and
this mineral when present is often changed into serpentine or into the
reddish-brown substance to which Lawson has given the name iddingsite. Hypersthene is present in some of the sections, and in others are
a few grains of quartz surrounded by augite borders.

Notes from Minnesota.—In a preliminary report of a season's
field work in northeastern Minnesota, Elftman[4] refers to the gabbro of
the region as producing contact metamorphism in the slates and schists
to the north of it. He describes more particularly the actinolite-mag-
netite slates, from near Birch Lake, that are believed to have originated in a fragmental rock whose nature, however, is not fully set forth.
The gabbro is an olivinitic variety. In it are great masses of anorthosite regarded by the author as phases of the gabbro. This is the rock

[4] 22d Ann. Rep. Geol. & Nat. Hist. Survey of Minn., p. 141.

which was reported by Lawson[5] as representing an old basement lying unconformable beneath the gabbro.

The Geology of Dartmoor, England.—McMahon[6] gives a few brief descriptive notes on some trachytes, felsites, mica-diorites, dolerites, tuffs and hornblende-schists from the western flank of Dartmoor. The trachytes and felsites are more or less altered, and the tuffs always very much so. The tuffs contain fragments of several kinds of lavas and of altered sedimentary rocks. The cementing material is " like the microgranular base of some rhyolites and porphyries." The most interesting rocks are the hornblende schists, which are thought by the author to be altered basic tuffs. They are marked by a fine grained parallelism of their constituents, producing a structure which the author designates the "corduroy structure." The rocks consist of augite, secondary hornblende and feldspar, the first two of which are often well crystallized. Their alteration is thought to be due to the intrusion of the tuffs by the great mass of epidiorite of the Cock's Tor. The basic schists of the Lizards that have been so repeatedly discussed, are believed to have had a similar origin.

Miscellaneous Notes.—The study of a series of nepheline rocks leads Gentil[7] to the conclusion that the peg structure so characteristic of this mineral is an effect of alteration. The alteration product is often a hydrated pleochroic substance with a yellowish tinge. The 'pegs' are produced by the extension of this substance along directions of feeble cohension in the original mineral (solution planes?)

The rock by whose decomposition the apophyllite[8] of Callo in Algeria was formed, is a biotite-augite-andesite, whose groundmass is usually more altered than the phenocrysts. The inclusions found in the rock are of cordierite gneiss, fragments of andalusite and of sillimanite and large segregations of plagioclase a little more basic than the feldspar of the phenocrysts.

On account of the similarity in crystalline structure between flint and Arkansas whetstone, Rutley[9] is inclined to regard the latter rock as derived by the replacement of limestone or dolomite by silica. The rhombohedral cavities noted by Griswold are thought to have been

[5] Bull. No. 8, Geol. & Nat. Hist. Survey of Minn.
[6] Quart. Jour. Geol. Soc., 1894, p. 338.
[7] Bull. Soc. Franc. d. Min., xvii, p. 108.
[8] Ib., p. 11.
[9] Quart. Jour. Geol. Soc., 1894, p. 377.

produced by the solution of some crystals of calcite or dolomite that remained for a time in the midst of the replacing silica after all the rest of the carbonate had been removed.

Pearce[10] gives a series of analyses to sustain his theory that the free gold of the Cripple Creek District, Colorado, has been mainly derived from the oxidation of tellurides.

In a preliminary report on the Rainy Lake Gold Region in Minnesota and Manitoba, H. V. Winchell and U. S. Grant[11] give some brief notes descriptive of the Laurentian, Coutchiching and Keewatin rocks of the district.

[10] Colo. Scient. Soc., April 5, 1894.
[11] 23d Geol. & Nat. Hist. Sur. of Minn., p. 36.

Reprinted from The American Naturalist, April 1st, 1895.

PETROGRAPHY.[1]

Granite Inclusions in Gabbro.—Inclusions of granite in the gabbro of the Cuillin Hills, Skye, England, afford excellent illustrations of the effects produced by the fusion of acid rocks on a molten basic one. The granite in question is reported by Judd[2] to be a biotite or a hornblende-biotite variety. Near the periphery of the mass the biotite and hornblende are replaced by augite, and granophyre is developed in the intersticee between the phenocrysts. The gabbro, in its passage upward, broke fragments from this granite, especially from its peripheral portions, and changed them completely. The granophyric intergrowth was fused and changed to a rhyolitic glass, marked by flow lines and filled with spherulites and lithophysae. In a few instances, some of the larger granophyre groups have escaped complete fusion, in which case, their remnants remain as nuclei of large compound spherulites. Imbedded in the glass are the large crystals of the granite. The quartzes have been cracked, and into the cracks glassy material has been pressed. The feldspars are also cracked, and in the crevices thus formed, secondary feldspars have been deposited. The original augites have disappeared, and in their places are aggregates of magnetite and other secondary products. The most interesting features of the altered inclusions are the spherulites. Simple and composite varieties are both common, and the trichitic kinds described by Cross are also met with. The centers of the spherulites are nearly always grains of quartz or of orthoclase, or groups of granophyre, as already mentioned. Pyrite and fayalite are both new products of the metamorphic action.

The Geology of Pretoria, South Africa.—A long and interesting account of the geology of the gold fields near Pretoria, in the

[1] Edited by Dr. W. S. Bayley, Colby University, Waterville, Maine.
[2] Quart. Jour. Geol. Soc., xlix, p. 175.

South African Republic, has appeared under Molengraaf's[3] name. The major portion of the paper is taken up with descriptions of the geological features of the region. There are in it, however, several items of petrographic interest. The oldest formation of the region embraces granites and crystalline schists. The former rock-type includes tonalites and orthoclase-plagioclase-microcline granites. In some places the rocks show evidences of dynamic metamorphism. Among the rocks associated with the granite are sericite-schists, actinolite-schists and amphibolites. Above these is another schist formation, comprising quartzites, clay slates, corundum-schists and porphyroids, and chiastolite-schists, cut by diabase dykes. The corundum porphyroid resembles a feldspar porphyry. Large crystals of biotite and large corundum individuals are in a groundmass of quartz and chlorite. The whole rock is besprinkled with quartz grains. Above the schists are bedded fragmentals, with which are associated diabases, quartz-porphyres and amygdaloids. In one of the diabases a diallagic augite and a primary hornblende were detected. In the carboniferous sediments south of Reitzburg are quartz gabbro and quartz diabases, and in the Rhenosterkop in the diamond fields at Driekop, in the Orange Free States, is a quartz-amphibole gabbro containing magnetite, biotite, primary hornblende, diallage and plagioclase. The pyroxene is striated parallell to oP, and is twinned parallel to ∞ P$\overline{\infty}$.

The Gabbro of the Adirondacks.—The gabbro associated with anorthosites of the Adirondacks are described by Smyth[4] as very similar to the Baltimore gabbros. They are best developed at Morehouseville and at Wilmurt Lake in the valley of West Canada Creek. The rock is a norite, in some phases a hypersthene-gabbro, both containing a brown hornblende regarded as original. The hypersthene, especially in the foliated varieties of the gabbro, which have been rendered schistose by pressure, sends tongues out into the contiguous feldspar. This stringing out of the pyroxene is so closely connected with the development of the foliation of the rock that it is believed to be a dynamic phenomenon. An analysis of the gabbro gave:

SiO_2	Al_2O_3	Fe_2O_3	FeO	MgO	CaO	Na_2O	K_2O	H_2O	Total
46.85	18.00	6.16	8.76	8.43	10.17	2.19	.09	.30 =	100.95

A black garnetiferous hornblende gneiss, which is associated with the other gneisses in the neighborhood of the gabbro, is thought to be re-

[3] Neues Jahrb. f. Miner. B. B., ix, p. 174.
[4] Amer. Jour. Sci., xlviii, 1894, p. 54.

lated to the latter rock, from which it is believed to have been derived by pressure. Around the garnets are rims composed of radiating tongues of hypersthene or of hornblende. Green hornblende is present in the gneiss in addition to the brown variety, and all the other components of the gabbro are represented in either the fresh or the altered condition.

The Dykes of the Thousand Islands.—The granites, gneisses and other rocks of the Admiralty Group of the Thousand Islands in the St. Lawrence River are cut by numerous dikes of a dark rock. These, to the number of thirty, have been studied by Smyth.[5] They are all normal diabases and olivine diabases. In the latter variety the olivine is often surrounded by a reaction rim composed of radiating plates of tremolite. The magnetite in many of the rocks of both varieties is separated from the plagioclase by a rim of biotite. This is absent when the mineral is in contact with the other rock components, hence it is regarded as a true reaction rim between the iron oxides and the feldspar.

Analcite-Diabases from California.—A series of dykes, from San Luis, Obispo Co., California, are described by Fairbanks[6] as consisting of two distinct portions. The main one is dark and fine-grained, and the other a hard, light, rock cutting the former in dykes. Both possess the same general features in the thin section, but the lighter rock possesses them in greater perfection. It consists of lath-shaped basic plagioclase, lamellar diallage and analcite. The latter mineral occurs as irregular masses in the feldspar, in wedge-shaped pieces between the plagioclase, in the form of hexagonal or rounded grains partly enclosed within the feldspars, and as the lining of cavities in the rock. It is supposed to have been derived from nepheline, as the mass analysis of the rock shows it to be very rich in sodium:

SiO_2	Al_2O_3	Fe_2O_3	FeO	CaO	MgO	K_2O	Na_2O	H_2O	Cl	Total
50.55	20.48	2.66	4.02	7.30	4.24	2.27	8.37	.44	tr	= 100.33

The analcite is changed partly to an aggregate of green fibres, and partly to natrolite. In the wedge-shaped areas between the plagioclase the mineral also contains prehnite crystals, and is bordered here and there by a doubly refracting substance supposed to be a soda feldspar. These are both believed to be alteration products of the analcite.

[5] Trans. N. Y. Acad. Sci., xiii, p. 209.
[6] Bull. Dept. Geol. Univ. Cal., Vol. I, p. 273.

In some of the dykes the structure is ophitic, and in others, panidiomorphic. If the author's view as to the origin of the analcite is correct, these rocks are clearly related to teschnites.

A Quartz-Keratophyre from Wisconsin.—Weidman[1] has investigated the porphyritic rock overlying the Baraboo quartzites of Wisconsin, and has shown it to be a quartz-keratophyre. It shows all the features of a lava, and is associated with tuffs and a sericite schist. The schist is at the contact of the keratophyre with the quartzite, and is evidently a result of shearing of the eruptive. The latter is porphyritic, with plagioclase and anorthoclase phenocrysts (often fractured by movements of the lava), and a few partially dissolved quartz phenocrysts in a fine-grained holocrystalline groundmass of quartz and feldspar, which, in addition to the phenocrysts mentioned, contains imbedded in it ilmenite, biotite and zircon. Many specimens show a fluxion structure and some are spherulitic—the spherules being sometimes secondary and sometimes primary bodies. An analysis of a sample of the rock gave:

SiO_2	Al_2O_3	FeO	CaO	K_2O	Na_2O	H_2O	SO_3	Total
73.00	15.61	1.95	.79	.88	4.95	1.06	.76	= 99.00

The series of bulletins, of which the author's article forms the second number, is well printed and is apparently well edited. It is a valued addition to the list of science bulletins now being published by American colleges.

Notes.—The crystalline limestones of Warren Co., N. J., contain a large number of accessory minerals, which are described by Westgate.[2] It contains irregular masses or concretions of pyroxene, hornblende, magnetite and biotite. Quartz, tourmaline, apatite, graphite and garnet are also present in it. The quartz and pyroxene are so abundant that, in some cases, they constitute rock-bodies, composed of interlocking grains of their principal constituents, with a small admixture of some others.

The nickeliferous pyrrohotite of the Gap Mine, Lancaster, Pa., forms a peripheral zone around the eastern end of an amphibolite lens, which, according to Kemp,[3] is an altered norite or peridotite. The ore is irregularly intermingled with the hornblende of the amphibolite,

[1] Bull. Univ. Wis. Science Ser., Vol. I, p. 85.
[2] Amer. Geologist, Vol. xiv, p. 308.
[3] Trans. Amer. Inst. Min. Engin., Oct.. 1894.

filling interstices between its crystals. The author is inclined to regard the ore as having separated from the rock magma, but, whether in accordance with the Soret principle, or not, he is unwilling to say.

A variolite in a small dyke at Dunmore Head, County Down, Ireland, is described by Cole[10] as an altered glass containing spherulites composed of cryptocrystalline material with a delicately radial structure. Cracks traverse the spherulites and also the groundmass of the rocks. Into some of those in the spherulites glass has been forced. Occasionally the nuclei of spherulites are crystals of plagioclase.

In a general geological article on the Essex and Willsboro' Townships in Essex Co., N. Y., White[11] records the existence of a number of bostonite, fourchite, camptonite and other dykes cutting the country rocks of the region.

Reprinted from The American Naturalist, May 1st, 1895.

PETROGRAPHY.[1]

The Eruptive Rocks of the Christiana Region.—Brögger[2] has done an excellent piece of work in this, the first of his reports on the eruptive rocks of Norway. The article deserves much more notice than can be given it in this place. Briefly, the author describes grorudite, salvsbergite and tinguite dykes which together form what is denominated a rock series—that is, a series of rocks that differ slightly from each other in their chemical composition, but which, at the same time, by their intimate gradations into each other, give evidence of being closely related. All of these rocks are rich in soda and potassa, and all contain alkaline amphiboloids. The grorudite is essentially an aggregate of microcline and albite, usually in microperthitic intergrowths, rarely anorthoclase, and always aegerine and amphibole, as phenocysts, in a groundmass of potash feldspar, albite, sometimes soda-orthoclase, aegerine and more or less quartz. The amphiboles are arfvedsonite and katoforite, the latter name being given to a series of alkaline iron amphiboles having the angle $C \wedge c = 31°-58°$, and pleochroism as follows: $B > C > A =$ yellowish red > brownish red > yellowish red or greenish yellow. In all their properties, so far as studied, they occupy a position between barkevikite and arfvedsonite. Salvsbergite differs from grorudite in containing little or no quartz. Its structure is trachytic.

Grorudite is regarded as the dyke form of soda-granite and pantellerite and salvsbergite that of nordmarkite.

After a discussion of the significance of the notion of dyke rocks as a group of well-defined rock types, the author concludes that while the group is well characterized by Rosenbusch, it includes a number of rocks that are but apophyses of bosses, etc., and which should be classed with the rocks of bosses. He prefers the term "hypabyssische Gesteine" for all rocks with the structure of dyke rocks, whether they be in the form of true dykes, of sheets, or whether they occur as the peripheral form of bosses or laccolites. The hypabyssmal rocks comprise a great group of equal value with that of the surface (volcanic) rocks and that of the abyssmal (plutonic) rocks. It includes two classes—the aschistic and the diaschistic—the first embracing those rocks not produced by the differentiation of their source magma, and the latter

---

[1] Edited by Dr. W. S. Bayley, Colby University, Waterville, Maine.
[2] Viedenskabsselskabels Skrifter. Math.-naturv. Klasse, 1894, No. 4.

those thus produced. The diaschistic rocks form complementary members, such as the minettes and aplites. The complementary form of salvsbergite is lindoite, a trachytic aggregate of phenocysts of microperthite and brown biotite, in a groundmass of quartz, biotite, aegerine and various secondary products, among which carbonates play an important *rôle*.

The laws of differentation in the different parts of the dykes are studied through the aid of a large number of carefully made analyses, as well as those governing the differentiation of the dyke masses from the boss masses. In all cases it is found that the differentiation consists in an increase in Fe_2O_3 toward the sides of the dyke, and an increase of the same constituents in the dyke masses as compared with the coroesponding boss material. The original magma is believed to have split into two magmas, one of which yielded the laccolite and boss material, and the other the substance of the diaschistic dykes. The former, in turn, split in the same way into a peripheral and a main phase, the former of which gave rise to the aschistic dykes.

The large number of analyses accompanying the discussion, and the careful description on which it is based, supply an excellent basis on which the long-desired genetic and philosophical classification of rocks may be founded, provided the lines of thought developed by the author are found to hold for other regions than those of southern Norway.

The Massive Rocks of Arran.—A very full account of the petrographical features of the massive rocks of the southern half of the Island of Arran has been given by Corstorphine[*]. The rock-types include pitchstones, quartz porphyries, normal diabase, quartzitic phases of the same rock, olivine-analcite varieties and sahlite diabases, all of which occur in sheets or dykes. The pitchstone presents no unusual characters. The quartz porphyries include those with a spherulitic groundmass and those whose groundmass is crystaline, and among the latter are microgranitic and micropegmantic varieties. The quartz-bearing diabases are usually in sheets. They contain large macroscopic quartzes and feldspars, especially near their contacts with the porphyry, and at their contacts with the underlying sandstone they contain large fragments of this rock. In the normal biabase both hypersthene and biotite occur. The large crystals of quartz and feldspar are regarded as foreign components, which have been caught up from the porphyry. The olivine analcite diabase is a typical diabase in which zeolites, and especially analcites, are abundant. These occupy the interstices be-

[*] Minn. u. Petrog. Mitth., XIV, p. 443.

tween the plagioclase and augites, and are thought to have originated
from the alteration of nepheline.

Migration of Crystals from a Younger to an Older Rock.

—It has long been assumed, that of two igneous rocks in contact, that
containing crystals peculiar to the other was necessarily younger than
the latter. Cole,[4] however, shows that crystals may be floated away
into a pre-existing rock of a low degree of fusibility from one of a
higher degree which has intruded it. At Glasdrumman Port, County
Down, Ireland, a dyke of eurite is flanked on both sides by dykes of
basaltic andesite, of which the andesites are unquestionably the older
rocks, since the eurite on its contact with them encloses fragments torn
from their sides. The eurite contains porphyritic crystals of pink
orthoclase, while the andesite is normally devoid of them. Near its
contact with the former rock, however, crystals exactly like those in the
eurite are occasionally found in the andesite. Crystals of quartz and
feldspar have also often been floated from the eurite into the detached
fragments of the andesite. The invading rock has melted the groundmass of the andesite and has left its larger crystals scattered through a
matrix made up largely of molten andesite intermingled with some
eurite substance.

Notes.—In a report accompanying an excellent geological map of
Essex Co., Mass., Sears[5] describes briefly the following rocks : Hornblende granitites, granophyric granitites with a flowage structure, augite-
nepheline syenites, hornblende diorites, quartz-augite-diorites, musco-
vite-biotite-granites, norites, quartz porphyries, peridotites, gneisses,
both igneous and clastic, bostonite and tinguaite dykes and various
effusive rocks.

A series of chemical analyses of the gneissoid granites, granite porphyries and porphyrites of the Bachergebirge in Stiermark, has been
made by Pontoni[6] in order to discover whether all the granite porphyries, that form great dyke masses in the region, have the same composition or not, and whether the small porphyrite dykes that cut the
granite are like the granites and the granite porphyries or are unlike
them. The conclusion reached is to the effect that the granite porphyries are identical with the gneissoid granites of the region, and that the
porphyrites are independent intrusives.

[4] Scient. Trans. Roy. Dub. Soc., Vol. V, Ser. II, p. 239.
[5] Bull. Essex Inst., XXVI, 1894.
[6] Min. u. Petrog. Mitth., XIV, p. 360.

Zaleski[7] has made, with great care, a number of chemical analyses and mechanical separations of several granites to determine whether or not they are syenites plus quartz ; that is, whether or not the chemical limits between which these rock types vary are fixed. His results may be tabulated as follows :

Locality.	SiO_2 Content.	SiO_2 of rock—Quartz.
Dannemora,	61.06	54.08
Nigg,	69.84	65.33
Hangö,	71.42	59.46
Baveno,	74.44	41.38

Of these granites only one possesses the silica content of syenite after the quartz has been abstracted from it.

Spurr,[8] in a bulletin on the iron-bearing rocks of the Mesabi Range in Minnesota, describes a series of fragmental and cherty rocks associated with the ores.   One of these, to which he gives the name "taconite," consists of a groundmass of silica, in which are granites of a green substance, regarded by the author as glauconite. These are always more or less altered, yielding siderite, magnetite, hematite, etc.   The sideritic phase of this taconite is like the original carbonate of Irving and Van Hise.

In a small collection of specimens from central and western Paraguay, Milch[9] has recognized quartzites, limestones and phonolites.

Reprinted from The American Naturalist, June 1st, 1895.

PETROGRAPHY.[1]

Rock Differentiation.—Harker[2] contributes an interesting article on rock differentiation in his study of the gabbro of Carrock Fell, England. The hill in question consists of bedded basic lavas, gabbro, granophyre and diabase in the order of their intrusion. The gabbro is of especial interest, since it presents a simple example of rock differentiation. In its center the mass is quartziferous. Toward the periphery it passes gradually into an ordinary gabbro, and immediately upon the border into an aggregate composed largely of titaniferous magnetite. In explaining the causes of this gradual transition in chemical and mineral composition, the author discards the theories usually proposed to explain similar phenomena, and concludes that, in the case under discussion, the separation of the magma into its parts took place during the period of crystallization by concentration of the crystallizing substances. The concentration is greatest for those minerals belonging to the earliest stages of the rock's history, hence it is thought that the differentiation took place by diffusion in a fluid magma, and that in those parts of this magma richest in basic minerals crystallization first occurred. As the crystals separated, the supply of the crystallizing substance was kept up by diffusion from other portions of the magma into the basic portions.

Another interesting feature of the gabbro mass relates to the contact effects produced by the rock in the surrounding basic lavas, some of which are enclosed as fragments in the midst of the gabbro. Their isotropic base has crystallized, and some changes have been produced

---

[1] Edited by Dr. W. S. Bayley, Colby University, Waterville, Maine.
[2] Quart. Journal Geol. Soc., 1894, p. 311.

in the composition and structure of their phenocysts. At the immediate contacts of the different rocks a commingling of their materials seems to have taken place. Mica has been generated in the gabbro, and the groundmass of the lavas has disappeared, leaving a plexus of small feldspar laths imbedded in a clear mosaic of quartz or of quartz and feldspar.

The Metamorphism of Inclusions in Volcanic Rocks.— In a memoir presented to the French Academy of Sciences, Lacroix[3] gives a very full resumé of the conclusions reached by him in the study of the action of modern volcanic rocks on the inclusions imbedded in them. The conclusions are based on the results of late studies as well as on those reached several years ago.[4] The author finds that the basaltic and the feldspathic effusives act differently toward foreign fragments imbedded in them. The former act principally through their high temperature, fusing the most easily melted components of the inclusions, while the trachytic rocks act more effectively in producing mineralogical changes through the aid of the mineralizers, mainly water, with which they are abundantly provided. The physical and chemical changes suffered by the material of the inclusions are discussed separately and fully. Often the fragments in the basalts are reduced by fusion to a few grains of their most resistant components, while the fragments in the trachytes have lost only their micaceous constituents by fusion. Consequently the metamorphism in the latter cases is supposed to have been produced at a comparatively low temperature, although the new minerals produced in number exceed by far those produced in the basaltic inclusions at a much higher temperature. With respect to the effects produced on rocks in situ, it is found that basaltic and trachytic lavas act alike—mainly through their heat. The metamorphic action in both cases is comparatively slight. The similarity in the effects produced by the two types of lavas in this case, when compared with the dissimilar effects produced upon their inclusions, is explained as a consequence of the fact that all lavas, when they reach the surface, lose their volatile constitutents, and so, of necessity, can affect alteration in contiguous rock masses solely by means of their high temperature. In other words, the alteration of inclusions is effected at a depth beneath the surface, while the alteration of rocks in situ is a surface phenomenon.

[3] Mèmoires présentés à l'Acad. d. Sciences de l'Institut de France, xxxi.
[4] See American Naturalist, 1894, p. 946.

The Petrography of Aegina and Methana.——The lavas of
the island of Aegina and the peninsula Methana in Greece are andesites and dacites that have broken through cretaceous and tertiary
limestones. Washington[5] separates the rocks into the two groups
above-mentioned on the basis of the SiO_2 contents. Rocks containing
above 62% of SiO_2 he classes as dacites, those containing less than this
amount as andesites. The dacites are divided into hornblende, horn-
blende-hypersthene and biotite varieties, and the andesites into hornblende, biotite-hornblende, hornblende-augite, hypersthene and horn-
blende-hypersthene varieties. All the rocks are more or less porphyritic,
and all contain more or less glass. Tridymite is present in the hornblende andesites from the Stavro district. The trachyte described by
Lepsius from near Poros is a biotite-hornblende-andesite. Brown and
green hornblendes are both present in the Grecian rocks, but not in the
same specimens. The green variety is characteristic of the pyroxene
free andesites, and the brown variety of those rocks containing an almost
colorless pyroxene as one of its essential components. This association
of the two hornblendes indicates that their formation is dependent upon
differences in chemical composition of the magmas from which they
separated, as well as upon the conditions under which their separation
took place.

In almost all of these rocks there are segregations of the same composition as that of the enclosing rocks, except that they are more basic.
Two classes of segregations are observed. The first are hornblende-
augite-andesites, containing brown hornblende and no glass; the second
class is composed of green hornblende in a glassy base with plagioclase
laths. The brown hornblendes are often changed to opacite, surrounded
by a zone of colorless crystals of augite. In those segregations in which
the hornblende is of the green variety, no such alteration is observable.
The glass in these segregations is so different from that of the rock in
which they occur, that it cannot be regarded as portions of the latter.
The author is inclined to regard these bodies as fragments of earlier
lava flows buried deeply beneath the latter ones.

In his discussion on the general relations of the different rocks of the
region, the author states that "in general　*　*　*　the more acid the
rock the more vitreous the groundmass, the smaller and more microlitic the crystals in it, and the larger and more abundant the phenocysts."　•

After remarks on the chemical relations of the different rock types
to each other, and a discussion of the Aegina-Nisyros region as a
"petrographical province," the paper closes with the statement that

[5] Jour. of Geology, Vol. II, p. 789, and Vol. III, p. 21.

although the lavas of the region under discussion are so similar to those of the Andes, nevertheless, the original undifferentiated magmas of the two districts were quite dissimilar.

Maryland Granites.—The granite and associated rocks on the east side of the Susquehanna River in Cecil County, Maryland, have been made the subject of study by Grimsley.[6] In the northern portion of the area investigated, the granite is but little sheared, while in its southern portion the rock is very gneissic. The two portions of the area are separated from each other by a band of staurolite-schist. Though the rocks of both areas were originally the same in composition, it is thought that the northern granite may be the younger, since it is intruded by dykes of what appears to be a dynamically metamorphosed gabbro, while, on the other hand, the southern granite intrudes a basic rock that apparently grades into gabbro. Both granites are biotitic varieties, and both are eruptive in origin. The northern granite is remarkable for the epidotization of its feldspar, which is predominantly plagioclastic, and for the occurrence in it of numerous dark basic segregations. Many rare minerals, such as zircon, magnetite, tourmaline, cubical garnets and sphene were found in large quantities in the soil produced by its decomposition. The northern contact of the northern granite is somewhat abnormal in its characters. The granite appears to become more basic toward the contact, and the basic phases are cut by apophyses of the normal acid rock.

An analyses of the granite follows :

SiO_2	TiO_2	Al_2O_3	Fe_2O_3	Feo	MnO	CaO	SrO	BaO	MgO	Na_2O	Li_2O	H_2O	P_2O_5	Total
66.68	.50	14.93	1.58	3.23	.10	4.89	tr.	.08	2.19	2.65	tr.	1.25	.10	= 100.32

Alabama Cherts.—Hovey[7] has recently examined a series of cherts sent him from Alabama. Those from the Lower Magnesian series consist almost entirely of chalcedony, with the addition of a little quartz and opal. The rocks are fine-grained mosaics that are mottled by reason of variations in the fineness of their grains. The quartz appears to be secondary, as it fills cavities in the chalcedony. A few scales of limonites and dust particles are present in almost all sections. No well-defined organic remains were detected in any. The cherts from the Lower Carboniferous, on the other hand, contain numerous remains of calcareous organisms, which are cemented together by chalcedony exhibiting a tendency to form concretionary granules. In some specimens, genuine spherocrystals of this mineral were detected. Chemical analysis of both classes of cherts show the absence of opal. The author regards the rocks as chemical precipitates.

[6] Jour. Cin. Soc. Nat. Hist., Apr.–July, 1894.
[7] Amer. Jour. Sci., 1894, xlvili, p. 401.

PETROGRAPHY.[1]

An Example of Rock Differentiation.—The Highwood Mountains of Montana have afforded Weed and Pirrson[2] an interesting study in rock differentiation. The mountains comprise a group of hills composed of cores of massive granular rocks surrounded by acid and basic lava flows and beds of tuff, which are cut by hundreds of dykes radiating from the cores as centers. One of these hills, isolated from the others is known as Square Butte, whose laccolitic origin can be plainly shown. The Butte is composed entirely of igneous rocks. Its center is a core of white syenite, and around this as a concentric envelope is a dark basic rock called by the authors shonkinite. Near the top of the Butte the surrounding envelope has been eroded off exposing the white rock, so that from a distance the latter appears to be capping the former. The black rock consists of biotite in large plates and augite crystals, in the irregular spaces between which are found orthoclase, olivine, a little albite and small quantities of nepheline, cancrinte and the usual accessory minerals. An analysis of the rock gave:

[1] Edited by Dr. W. S. Bayley, Colby University, Waterville, Maine.
[2] Bull. Geol. Soc. Amer., Vol. 6, p. 389.

SiO$_2$	TiO$_2$	Al$_2$O$_3$	Fe$_2$O$_3$	FeO	MnO	MgO	CaO	Na$_2$O	K$_2$O	H$_2$O	P$_2$O$_5$	Cl	Total
46.73	.78	10.05	3.53	8.20	.28	9.68	13.22	1.81	3.76	1.24	1.51	.18	=100.97

The rock is thus a granular plutonic rock consisting essentially of augite and orthoclase. It is closely related to augite-syenite, bearing the same relation to it as vogesite does to hornblende-syenite.

The white rock associated with the shonkinite is a sodalite-syenite, containing as its bisilicate component only amphibole. Its composition is given as follows :

SiO$_2$	TiO$_2$	Al$_2$O$_3$	Fe$_2$O$_3$	FeO	MnO	MgO	CaO	Na$_2$O	K$_2$O	H$_2$O	P$_2$O$_5$	Cl	Total
56.45	.29	20.08	1.31	4.39	.09	.63	2.14	5.61	7.13	1.77	.13	.43	=100.45

The basic rock is richer in iron, magnesia and lime than the acid one ; since the two rocks pass into each other by a rapid but continuous gradation, they are believed to be of the same age and to be the complementary differentiated portions of the same magma. The differentiation in this case could not have been due to a process of crystallization, in which the first crystallized minerals were accumulated in the peripheral portions of the cooling magma, since the other iron-bearing components of the shonkinite and of the syenite are so radically different. The differentiation must have occurred in the magma while still molten.

The Serpentines of the Central Alps.—Three years ago Weinschenck[3] gave a preliminary account of the serpentines of the East Central Alps and their contact effects, showing that the former were originally pyroxene eruptives. In a recent paper he returns to the subject,[4] and in a well illustrated article gives in detail the reasons for his former conclusions. He finds upon the examination of a large suite of specimens that the original rock was an olivine-antigorite aggregate, which he names stubachite, from its most important locality. The antigorite is believed to be an original component and not an alteration product of the olivine, as it is found intergrown with perfectly fresh grains of the latter mineral. The grate structure (" Gitterstructur ") of many serpentines is ascribed to such intergrowths, and not to the alteration of pyroxene along its cleavage planes. The original stubachite was a medium grained holocrystalline, allotriomorphic rock of intrusive igneous origin, which has not suffered much alteration since its exposure by erosion.

[3] American Naturalist, 1892, p. 767.

[4] Abhand. d. k. bayer. Ak. d. Wis II, Cl. XVIII, Bd. p. 653.

Becke[5] calls attention to the frequency with which a pyroxenic origin has been ascribed to serpentines of the Alps because of the lack in them of the mesh structure, and questions the safety of this conclusion when based on such scanty premises. He mentions the existence of a serpentine in the stubachthal in the Central Alps, in the freshest portions of which olivine and picotite can be seen in large quantities, and in other portions diopside and olivine. In many specimens the olivine has been crushed into a mosaic, the finer grains of which have been altered into serpentine, clinochlor, antigorite and what is probably colorless pyroxene. The mesh structure is found in the weathered portion of the antigorite-serpentine. It is thought by the author to be due to weathering subsequent to the production of the antigorite.

The central mass of the east central Alps consists of granite and gneiss,[6] of which the former is intrusive in the latter, although both have essentially the same mineralogical composition, and the former is schistose on its periphery. The granite contains zoisite, epidote, orthite, chlorite, calcite, etc., all of which are regarded as original, since the other primary components of the rock from which they may be assumed to have come are perfectly fresh. The origin of these minerals is ascribed to the cooling of the magma under the influence of mountain-making processes—a condition of crystallization which the author designates as piezocrystallization. The hydrated components of the rock are supposed to have been formed with the aid of magma moisture under the influence of pressure. This theory is believed to account for the granulation and other pressure phenomena noted in the granite, as well as for its composition.

Dynamic Metamorphism.—In connection with his work on the rocks of the Verrucano in the Alps, Milch[7] makes a study of dynamic metamorphism and suggests a number of terms to be used in the descriptions of metamorphic rocks. Allothimorphic fragments are those with the composition and forms of the original grains. Authimorphic fragments have the forms of the grains changed but their composition unchanged. Allothimorphic pseudomorphs have the original forms but a composition different from that of the original grains, and authimorphic pseudomorphs have both forms and composition changed, but with the latter dependent upon the original composition. Finally eleutheromorphic new products are those entirely independent of the

[5] Minn. u. Petrog. Mitth., XIV, 1894, p. 271.

[6] Ib., p. 717.

[7] Neues Jahrb. f. Min., etc., IX, p. 101.

original substances both in form and composition. Of the authimorphic fragments two classes are noted, first, the authiclastic, those that have been unable to adapt themselves to the altered conditions and, consequently, which have been fractured, and, second, the kamptomorphic, embracing those fragments that have been able to adapt themselves to changed conditions, and so have yielded to these and have bent, or have assumed abnormal optical properties, such as undulous extinctions. With these terms the author describes some of the rocks studied and states that in many instances no traces of clastic structure remain in them, although they must be regarded as regionally metamorphosed fragmentals. Regional metamorphism, he declares, may be brought about by pressure alone, or by dislocation—pressure with movement (dynamic metamorphism). The former may act slowly, deforming the minerals in rocks, while the latter acts rapidly, shattering them. The latter process usually forms rocks like the mica-schists, with a fine grain, and the former coarse grained ones like the gneisses. Of course, the action of water, which is the agent of transportation of the new substances added during metamorphism, may come into play in each case. The Verrucano rocks exhibit the effects of both kinds of regional metamorphism. The article contains a great many suggestions of interest to students of metamorphism.

Miscellaneous.—The conglomerates and albite schists of Hoosac Mountain, Mass., referred[8] to some time ago in these notes, have been described by Wolff[9] in some detail in his report on the geology of Hoosac Mountain. The conglomerates form gneisses which grade upward into the albite schists. Amphibolites also are described, whose origin is from a basic intrusive rock. A large number of photographs of hand specimens and thin sections of the rocks described accompany the paper.

Van Hise[10] in the report by Irving and himself on the Penokee iron district, gives a number of descriptions of sedimentary and volcanic rocks, illustrated by a large number of plates of thin sections. The rocks discussed include greenstone conglomerates, crystalline schists, intrusive greenstones, slates, quartzites, limestones, etc.

Ries[11] finds that one of the crystalline schists of the series of foliated rocks forming the greater portion of Westchester Co., N. Y., is a

[8] American Naturalist, 1892, p. 768.

[9] Min. XXIII, U. S. Geol. Survey, p. 41.

[10] Mon. XIX, U. S. Geol. Survey.

[11] Trans. N. Y. Acad. Sci., Vol. XIV, p. 80.

plagioclase-augen-gneiss which the author calls a schistose granite-diorite. Its constituents are quartz, plagioclase, biotite, hornblende and orthoclase as its principal components, with garnet, sphene, zircon, apatite, muscovite and microcline as the accessories. The quartz is penetrated by rutile needles. Nearly all the rock's constituents show evidence of dynamic fracturing.

Reprinted from The American Naturalist, August 1st, 1895.

PETROGRAPHY.[1]

The Rocks of Gouverneur, N. Y.—An interesting feature of the biotite-hornblende gneisses[2] of the vicinity of Gouverneur, N. Y., is

[1] Edited by Dr. W. S. Bayley, Colby University, Waterville, Me.
[2] C. H. Smyth, Jr., Trans. N. Y. Acad. Sciences, xii, p. 203.

the abundance in them of microperthitic intergrowths of orthoclase and
plagioclase. From the relations of the plagioclase to the orthoclase
and to the surrounding minerals there can be no doubt that it is of
secondary origin. It fills cracks between quartz and orthoclase, and
from these areas it sends long stringers into the orthoclase along its
cleavage cracks and into its fracture lines, without suffering the least
interruption in its continuity. The gneiss in its structure is sometimes
granular and sometimes granulitic, and in the appearance of its constituents it shows plainly that it is a dynamo-metamorphosed rock. The
dark bands occurring with the predominating light colored ones consist, as a rule, of the same minerals as the latter, but one band noted is
composed of monoclinic pyroxene and hornblende in addition to the
feldspars. The normal granites of the region differs in composition from
the gneiss in the absence from them of hornblende, except in certain
basic segregations. The granite, like the gneiss, has suffered the effects
of pressure, but to a more limited extent. Among the limestones
associated with these rocks are phases containing much colorless pyroxene, tremolite and scapolite. Near the base of the limestone series
the pyroxene-scapolite rocks are foliated, and are apparently interstratified with unaltered beds. They consist of feldspar, quartz, pyroxene,
mica, sphene, apatite, graphite, pyrrhotite and pyrite, or of these components, with the feldspars replaced by secondary scapolite.

Diorites and Gabbro at St. John, N. B.—Among the intrusive rocks cutting the Laurentian near St. John. N. B., Matthew[*]
finds a granite-diorite and a gabbro. The diorite is coarse grained and
porphyritic in its larger masses, and fine grained and granular in its
smaller bands. Quartz, plagioclase, orthoclase, hornblende, biotite and
the usual accessory constituents compose the rock, while epidote and
microcline-microperthite are present in it as alteration products of
plagioclase and orthoclase. The microperthite is also noted as forming
a rim between plagioclase and quartz. As the rock becomes finer
grained orthoclase and biotite diminish in quantity. Although the
contacts of the diorite with the surrounding rocks are usually faulted,
it can be clearly seen that the latter have been altered by the intrusive.
On the contact with a gabbro, this latter rock has been changed to a
granular aggregate of hornblende and plagioclase. The diorite, on the
other hand, is very fine grained, and is composed of an allotriomorphic
mixture of plagioclase, quartz, orthoclase and a few small shreds and
grains of hornblende and biotite. Limestone in contact with the

[*] Trans. N. Y. Acad. Sci., xiii, p. 185.

eruptive has been marbleized. In it are pyroxenes and garnets, the latter often in large numbers. This diorite has heretofore been regarded as a metamorphosed sediment, but, from the evidence at hand, the author concludes that it is a true irruptive. The gabbro of the region is confined to two small knobs. In one, the rock grades from an anorthosite into a peridotite. In the latter phase olivine constitutes nearly half of its mass. Hypersthene is abundant, while augite, plagioclase, and the usual accessories, spinel and magnetite, are present in small quantities. Reactionary rims always surround the olivines when in contact with plagioclase. These are composed of three zones, an inner one of hypersthene which is continuous with the large hypersthene components; a middle one, composed of fine needles of uralitic amphibole, and an outer zone consisting of uralite and a deep green, highly refracting substance in grains, probably a spinel. The contact rim is supposed to be secondary. The various phases of the rock are usually much altered into actinolitic varieties.

South American Volcanics.—The collection of Argentine volcanic rocks belonging to Berlin University has been investigated by Siepert.[4] The collection embraces quartz-porphyries, porphyrites, diabases, augite-porphyrites, melophyres and an epidiorite-porphyrite. In the quartz-porphyries quartz grains are often surrounded by aureoles of the same substance, whose optical orientation coincides with that of the surrounded particles. Many of the grains show undulous extinction, which the author regards as secondary. In some of the specimens the granophyric structure, in others the microgranitic, and in still others the felsophyric structure predominates. In many instances the granophyric structure is unquestionably secondary. The porphyrites include diorite-porphyrite, eustatite-porphyrite and epidiorite-porphyrite. In one of tha latter a feldspar granule was seen to be surrounded by a feldspar aureole. The other rocks examined present no unusual features.

Specimens of the younger volcanic rocks gathered by Sapper in Guatemala were submitted to Bergeat[5] for study. They comprise trachytes, rhyolites, dacites, andesites and basalts. The trachytes, though of the "Drachenfels" type, contain about 66 % of silicia, and are thus closely related to the rhyolites. The andesites are the most abundant types. They include pyroxene, hornblendic and mica hornblende varieties. Some of the pyroxenic andesites contain two pyrox-

[4] Neues Jahrb. f. Min., etc., B. B., ix, p. 393.
[5] Zeits. d. deutsch. geol. Ges. xlvi, I, p. 126.

enes—a hypersthene and an augite, both of which are pleochroic in the same tints parallel to *B* and *C*, a difference of color being noticeable only in the direction of *A*. The author notes that the volcanoes on the principal fissures have eruptive andesites, while the others have yielded basalts.

Rock Classification.—A new classification of in organic rocks, based on the nature and past history of their components, is proposed by Milch.[6] The original rocks are the *archaiomorphic*, embracing those whose constituents have separated from a molten magma. Through alteration processes these have given rise to the *neomorphic* rocks, including the three groups: *anthi-lytomorphic*, *allothi-stereomorphic* and *anthi-neomorphic*. The first of these groups includes those rocks whose material was originally in some other condition, but whose constituents possess forms independent of outside influences, as, for instance, the chemical precipitates. The second group embraces those whose material has been transported and been laid down with its own form to produce a rock different from the original one, as the mechanical sediments. The third group comprehends rocks whose material is in its original position, but in a different condition from the original one, as in the case of the residual and metamorphic rocks.

Miscellaneous.—Levy and Lacroix[7] describe a Carboniferous leucite-tephrite from Clermain, that is associated with micaceous porphyrites. The tephrite contains large leucites and pyroxenes in a groundmass composed of biotite, augite, plagioclase and leucite. All of this latter mineral, whether in large or small crystals, is transformed into aggregates of albite.

Palache[8] announces the discovery of riebeckite and aegerine in the

[8] Neues Jahrb. f. Min., etc., 1895, I, p. 100.

Forellen granulite of the Gloggnitzer Berges, near Wiener-Neustadt in Austria. The rock is a typical granulite, consisting of a quartz-plagioclase aggregate in which are imbedded acicular crystals and grains of the amphiboloids mentioned.

[6] Neues Jahrb. f. Min., etc., B. B., ix, p. 129.
[7] Bull. Soc. Franc. d. Min., xviii, p. 24.

General Notes.

PETROGRAPHY.[1]

The Lherzolites of the Pyrenees and their Contact Action.—The contact action of the lherzolites of the Pyrenees upon the lower Jurassic rocks through which they cut has been studied carefully by Lacroix,[2] who publishes his conclusions in a volume illustrated

[1] Edited by Dr. W. S. Bayley, Colby University, Waterville, Me.

[2] Comptes Rendus, Feb. 11, 1895. Nouv. Archiv. D'hist. Nat., III, Sér. vi, p. 209.

by six plates containing fifty figures. The intensity of the metamorphism varies widely. At 500 meters from the contact the limestones are filled with metamorphic minerals, and even at 1.5 kilos from the nearest visible contact with the eruptive the limestones still contain many of these. The altered sedimentary rocks are limestones, calcareous marls and occasionally sandstones. In the limestones the principal new minerals found are dipyr, micas, feldspars, tourmaline, rutile, sphene, magnetite, hematite, pyrite, apatite, quartz, graphite and rarely spinel, epidote and garnet. ·The calcareous marls have been changed to aggregates of silicates with four types of structure, the honestone, the micaceous schist and the amphibolitic and dioritic. Near the contact the organic coloring matter of the marls has disappeared. A little further away it is changed to graphite and at a greater distance it remains intact. The fissures cutting through the metamorphic rocks are lined with zeolites, which, however, the author does not think are connected in any way with the metamorphic processes. The sandstones, at the only contact seen, were changed into quartzites rich in needles of rutile, and a lusite, sillimanite and a few flakes of mica. A close similarity exists between the contact action of lherzolites and granites. The difference in the two cases consists in a corrosion of the metamorphic rocks by the granite and a great production of feldspar, while in the case of the lherzolites there is no transition between the metamorphosing and the metamorphosed rocks. The conditions determining the nature of the contact rock formed are: 1, the original composition of the sedimentary beds; 2, the quantity of the volatile and soluble substance accompanying the eruptive; and 3, the conditions under which the rock was erupted.

Nepheline Rocks from the Kola Peninsula.—A full account of the nepheline syenite region of the Kola Peninsula, Finland, by Ramsey[2] and Harkman has recently appeared. The main results of the senior author's study of the region have already been given in these notes. Other results can only be referred to, as they are two numerous to be described in detail. The authors define a new rock type—imandrite. It is a rock composed of quartz, plagioclase, chlorite, biotite and several accessory components. The first two minerals occur in isometric grains separated from each other by seams of chlorite or biotite. The rock has a half clastic structure, since the quartz and feldspar appear often as fragments in the interstitial chlorite. The quartz is largely secondary, and is supposed to be due to a silicification

[2] Fennia, 11, No. 2, 1894. Also American Naturalist, 1892, p. 334.

of the original rock. A second type of imandrite resembles a silicified porphyritic rock. A hypersthene-cordierite-hornfels, with handsome cordierite crystals, an oliving-actinolite schist, containing cordierite, and several contact metamorphosed sediments are .described in detail. The major portion of the article deals with the nepheline syenites and the related rocks—theralites, augite,-porphyrites, iolites, monchiquites, tinguaites, etc., and the new rocks, lujavrite and tawaite. The theralite agrees exactly with Rosenbusch's definition of the type. It is a medium grained aggregate of idiomorphic pyroxene, and granitic plagioclase and nepheline, with the accessories brown hornblende, biotite, sphene, magnetite, apatite, sodalite and secondary zeolites. Lujavrite is a trachytic nepheline-syenite with its components largely idiomorphic. Tawaite is a coarse-grained mixture of sodalite and pyroxene.

Around the periphery of the nepheline syenite the rock is different from its main mass and it has produced contact effects with surrounding rocks. A nepheline syenite with a trachytic structure is described among the peripheral phases of the syenite, and a rock resembling pulaskite, but containing no porphyritic crystals. ˙This rock, which the authors call umptekite, is a nepheline syenite, poor in nepheline. It differs from the nepheline syenite in containing a calcium-feldspar, from augite-syenite in possessing hornblende instead of augite, from laurvikite in its structure, and from akerite in its lack of quartz. Its structure is granitic. Arfvedsonite is its principal amphiboloid, and besides, it possesses aegerine. The characteristic minerals of the nepheline syenite are also present in it. The aegerine is frequently associated with sodalite or with feldspar in pegmatitic intergrowths. A sillimanite gneiss is mentioned as possibly being a metamorphized sediment.

The Matrix of Naxos Corundum.—The corundum[4] of Naxos occurs in an iron gray foliated or massive granular rock composed almost exclusively of corundum and magnetite. The first mentioned mineral is in largest quantity. Associated with these two components are limonitic and hematitic alteration forms of magnetite, margarite, tourmaline, muscovite, cyanite, staurolite, biotite, rutile and occasionally spinel, vesuvianite and pyrite. The corundum is in rounded grains or in well defined crystals surrounded by magnetite. Most of the other constituents, with the exception of the magnetite, appear to be the results of shearing. An analysis of the rock gave: Corundum

[4] Tschermak, Min. u. Petrog. Mitth., xiv, p. 311.

PARTIAL INDEX OF SUBJECTS.

INDEX OF AUTHORS.

A

SUMMARY OF PROGRESS

IN

PETROGRAPHY

IN

1896.

BY

W. S. BAYLEY.

FROM MONTHLY NOTES IN THE "AMERICAN NATURALIST."

PRICE 50 CENTS.

WATERVILLE, ME.:
PRINTED AT THE MAIL OFFICE.
1897.

No. 15.

A

SUMMARY OF PROGRESS

IN

PETROGRAPHY

IN

1896.

BY

W. S. BAYLEY.

FROM MONTHLY NOTES IN THE "AMERICAN NATURALIST."

PRICE 50 CENTS.

WATERVILLE, ME.:
PRINTED AT THE MAIL OFFICE.
1897.

General Notes.

PETROGRAPHY.[1]

The Origin of Adinoles.—Hutchings[2] has discovered a contact rock at the Whin Sill, England, which, in the author's opinion, represents an intermediate stage in the production of an adinole from a fragmental rock. It contains corroded clastic grains of quartz and feldspar in an isotropic base containing newly crystallized grains of quartz and feldspar. The isotropic material is derived from the clastic grains by the processes of contact metamorphism, whatever they may be, as grains of quartz are often seen with portions of their masses replaced by the substance. The rock has begun its recrystallization from the isotropic material produced by solution or fusion of the original grains, but the process was arrested before the crystallization was completed. The paper concludes with some general remarks on metamorphism. The author thinks that the statement that in granite contacts no transfer of material takes place has not yet been proven true. He also thinks that more care should be taken in ascribing to dynamic metamorphism certain effects that may easily be due to the contact action of unexposed dioritic or granitic masses.

Notes from the Adirondacks.—The limestones, gneisses and igneous intrusives of the Northwestern Adirondack region are well described by Smyth.[3] The intrusions consist of granites, diorites, gabbros and diabases. The gabbro of Pitcairn varies widely in its structure and composition, from a coarse basic or a coarse, almost pure feldspathic rock to a fine grained one with the typical gabbroitic habit.

[1] Edited by Dr. W. S. Bayley, Colby University, Waterville, Me.
[2] Geological Magazine, March and April, 1895.
[3] Bull. Geol. Soc. Amer., Vol. 6, p. 263.

Compact hornblende is noted as an alteration product of its augite.
Where in contact with the limestones the gabbro has changed these
rocks into masses of green pyroxene, garnet, scapolite and sphene. A
second variety of the gabbro is hypersthenic. A third variety is characterized by its large zonal feldspars composed of cores of plagioclase
surrounded by microperthite, although crystals of the latter substance
alone abound in some sections. The ferromagnesian components are
rare as compared with the feldspars. Nearly all specimens of these
rocks are schistose, and all of the schistose varieties exhibit the cataclastic structure in perfection. Analysis of the normal (I) and of the
microperthitic or acid (II) gabbros yielded :

	SiO$_2$	Al$_2$O$_3$	FeO	MgO	CaO	K$_2$O	Na$_2$O	H$_2$O	Total
I	57.00	16.01	10.30	1.62	6.20	3.53	4.35	.15 = 99.16
II	65.65	16.84	4.01	.13	2.47	5.04	5.27	.30 = 99.71

Near the contact with the limestone the gabbro is finer grained than
elsewhere. Pyroxene is in larger grains than in the normal rock, but
the feldspar is in smaller ones. The limestone loses its banding and is
bleached to a pure white color. Between the two rocks is a fibrous
zone of green pyroxene and wollastonite, together with small quantities of
sphene and garnet and sometimes scapolite and feldspar. The red
gneisses, common to that portion of the region studied which borders
on the gabbro, are thought by the author to be largely modified portions of the intrusive rock.

The Eastern Adirondacks have been studied by Kemp.[4] The limestones of Port Henry consist of pure calcite, scattered through which
are small scales of graphite, phlogopite and occasionally quartz grains,
apatite and coccolite. This is cut by stringers of silicates that are
granitic aggregates of plagioclase, quartz, hornblende and a host of
other minerals. Ophicalcite masses are also disseminated through the
limestones, and these are also penetrated by the silicate stringers.
Merrill[5] has shown that the serpentine of the ophicalcite is derived
from a colorless pyroxene. The schists associated with the limestones
are briefly characterized by the author. At Keene Center a granulite
was found on the contact of the ophicalcite with anorthosite.

Hornblende Granite and Limestones of Orange Co., N. Y.
—Portions of Mts. Adam and Eve at Warwick, Orange Co., N. Y., are
composed of basic hornblende granite that is in contact with the white

[4] Ibid, p. 241.
[5] Cf. AMERICAN NATURALIST, 1895, p. 1005.

limestone whose relations to the blue limestone of the same region have been so much discussed. The granite contains black hornblende, a little biotite, and so much plagioclase that some phases of it might well be called a quartzdiorite. Allanite and fluerite are also present in the rock, the former often quite abundantly. As the granite approaches the limestone it becomes more basic. Malacolite, scapolite and sphene are developed in it in such quantity, that immediately upon the contact the normal components of the granites are completely replaced. On the limestone side of the contact the rock becomes charged with silicates, the most abundant of which are hornblende, phlogopite, light green pyroxenes, sphene, spinel, chondrodite, vesuvianite, etc. The contact effects are similar in character to those between plutonic rocks and limestones elsewhere. The blue and the white limestones are regarded as the same rock, the latter variety being the metamorphosed phase.[6]

An Augengneiss from the Zillerthal.—The change of a granite porphyry into augengneiss is the subject of a recent article by Fütterer.[7] The rocks are from the Zillerthal in the Alps. The gneisses are crushed and shattered by dynamic forces until most of the evidences of their origin have disappeared. The original phenocrysts have been broken and have suffered trituration on their edges, while new feldspar, quartz, malacolite and other minerals have been formed in abundance. The groundmass of the gneiss is a mosaic whose structure is partially clastic through the fracture of the original components and partially crystalline through the production of new substances. The author's study is critical, and, though he treats the described rocks from no new point of view, he discusses them with great thoroughness, calling attention at the same time to the important diagnostic features of dynamically metamorphosed rocks.

Petrographical News.—Ransome[8] has discovered a new mineral, constituting an important component of a schist occurring in the Tiburon Peninsula, Marin Co., Cal. The other components of the schist are pale epidote, actinolite, glaucophane and red garnets. The new mineral, lawsonite, is orthorhombic with an axial ratio .6652 : 1 : .7385, a hardness of 8 and a density 3.084. The axial angle is $2V = 84° 6'$ for sodium light. Its symbol is $H_4 Ca Al_2 Si_2O_{10}$.

[6] J. F. Kemp and Arthur Hollick: N. Y. Acad. Sci., VII, p. 638.
[7] Neues Jahrb. f. Min., etc., B.B. IX, p. 509.
[8] Bull. Geol. Soc. Amer., Vol. 1, p. 301.

Fuess[9] has perfected an attachment for the microscope which enables an observer to enclose with a diamond scratch any given spot in a thin section, so that it may be easily identified for further study.

Marsters[10] describes two camptonite dykes cutting white crystalline limestones near Danbyborough, Vt. They differ from the typical camptonite in being much more feldspathic than the latter rock. They moreover, contain but one generation of hornblende, corresponding to the second generation in the typical rock, and but few well developed augite phenocrysts, although this mineral is found in two generations.

A portion of Mte. S. Angelo in Lipari consists of a porous yellowish pyroxeneandesite containing grains and partially fused crystals of cordierite, red garnets and dark green spinel.[11]

Cole[12] declares that the "hullite" described by Hardman as an isotropic mineral occurring in the glassy basalts of Co. Antrim, Ireland, is in reality an altered portion of the rock's groundmass, and is no definite mineral substance.

The same author[13] describes the old volcanoes of Tardree in Co. Antrim as having produced rhyolitic lavas instead of trachytic ones as has generally been stated.

Reprinted from The American Naturalist, January 1st, 1896.

General Notes.

PETROGRAPHY.[1]

Igneous Rocks of St. John, N. B.—W. N. Mathew has con.
tinued his work on the igneous rocks of St. John, N. B.,[2] contributing
in a recent article an account of the effusive and dyke rocks of the
region.   All the rocks described are believed to be pre-Cambrian in
age.   They embrace quartz-porphyries, felsites, porphyries, diabases
and feldspar-porphyrites among the effusive rocks, and dierite-porphy-
rites, diabases and augite-porphyrites among the dyke forms.   In some
of the quartz-porphyries perlitic cracks may still be recognized, and in
the felsite porphyries some spherulites.   Tuffs of all the effusives are
abundant.   A soda granite with augite and green hornblende and
probably a little glaucophane was also met with.   It is intrusive, and
has a composition represented by the figures :

SiO_2	TiO_2	Al_2O_3	Fe_2O_3	FeO	MnO	CaO	MgO	Na_2O	K_2O	CO_2	Loss
64.86	.70	15.02	5.53	1.01	.18	2.61	1.42	3.92	2.37	.55	1.73

[1] Edited by Dr. W. S. Bayley, Colby University, Waterville, Me.
[2] Trans. N. Y. Acad. Science, XIV, p. 187.

The diorite-porphyrite has a groundmass of idiomorphic hornblende, lathshaped feldspars and some interstitial quartz, with phenocrysts of the same minerals, but principally of feldspar. Among the diabases is a quartzose variety.

Eruptive Rocks from Montana.—Among some specimens of eruptive rocks obtained from Gallatin, Jefferson and Madison Counties, Montana, Merrill[3] finds basalts, andesites, lamprophyres, syenites, porphyrites, wehrlites, harzburgites and websterites, some of which possess peculiar characteristics. A hornblende andesite, for instance, contains large corroded brickred pleochroic apatite crystals, whose color is due to innumerable inclusions scattered through them. The groundmass of some of the basalts has a spherulitic structure. The wehrlite is a holocrystalline aggregate of pale green diallage, reddish brown biotite, colorless olivine and a few patches of plagioclase. Its structure is cataclastic or granulitic, the larger crystals being surrounded by an aggregate of smaller ones. The websterite consists of green diallage and colorless enstatite with included foliae of mica and occasional interstitial areas of feldspar, and is thus related to gabbro. Some of the lamprophyres are composed of groups of polysomatic olivines or of olivine and au ite in a scaly granular groundmass of lighter colored minerals, through which are scattered small flakes of brown biotite and tiny augite microlites. This structure is accounted for on the supposition that the granular groups of olivine and of olivine and augite belong to an older series of crystalline products than those of the groundmass.

Porphyrites and the Porphyritic Structure.—In a general account of the laccolitic mountains of Colorado, Utah and Arizona, Cross[4] gives a brief synopsis of the characteristics of the rocks that constitute their cores. These rocks comprise augite, hornblende and hornblende mica-porphyrites, diorites and quartz-porphyrites. All contain phenocrysts of plagioclase and of the iron bearing silicates, with the feldspars largely predominating. These upon separating left for consolidation into the groundmass a magma which upon crystallization yielded a granular aggregate consisting largely of quartz and orthoclase. No pressure effects were seen in any of the sections studied. All are porphyritic with a granular groundmass, which differs in the different rocks, principally in the proportion of its constituents. The porphyritic structure as defined by the author is not the result of the recur-

[3] Proc. U. S. Nat. Museum, XVII, p. 637.
[4] 14th Ann. Rep. U. S. Geol. Survey.

rence of crystallization, producing several generations of crystals, but
it is a structure exhibiting contrasts in the size and form of the component crystals of a rock, resulting from the differences in conditions
under which the different minerals crystallized.

Granophyre of Carrock Fell, England.—In the Carrock Fell
district is a red granophyre closely associated with the gabbros. This
rock has recently been studied by Harker,[5] who had previously investigated the gabbros. The normal type of the granophyre is an augitic
variety in which the augite occurs as a deep green species which is idiomorphic toward the feldspars. Oligoclase is also present as idiomorphic
crystals in a reddish quartz-feldspar groundmass with the typical granophyric structure. The composition of the rocks is represented as follows:

SiO_2	Al_2O_3	Fe_2O_3	MgO	CaO	Na_2O	K_2O	Loss	Total
71.60	13.60	2.40	.21	2.30	5.55	3.53	.70	= 99.89

As the rock approaches the gabbro it becomes less acid and the proportion of augite in it increases. This is the lower portion of the mass
as it was originally intruded. Its more basic nature as compared with
the rest of the rock is explained as due to the absorption of parts of the
gabbro with which the granophyre is in contact.

The same author[6] also records the existence of a greisen, which is a
phase of the well known Skiddau granite. The greisen consists essentially of quartz and muscovite, but remnants of orthoclose are still to
be detected in it. The mica is regarded as having been derived largely
from the feldspar.

Sheet and Neck Basalts in the Lausitz.—The basalts of the
neighborhood of Seifeirnersdorf and Warnsdorf in the Lausitz, Saxony,
occurs in sheets according to Hazard,[7] and in volcanic rocks. The
sheet rocks are nepheline basalts, nepheline basanites and feldspathic
glass basalts. The neck forms are hornblende basalts, sometimes with
and sometimes without nepheline. The constituents of all are magnetite, apatite, augite, biotite, nepheline and glass in varying quantities,
with feldspar, olivine and hornblende in different phases. Sometimes
the mineral nepheline is absent, but this happens mainly in the glassy
varieties, where its components are to be found in the glassy base.
There are intermediate varieties between the hornblende and the olivine basalts corresponding to geological masses intermediate in characteristics between volcanic sheets and necks. In many of the neck

[5] Quart. Journ. Geol. Soc., 1895, p. 125.
[6] Ibid, p. 139.
[7] Min. u. Petrog. Mitth., XIV, p. 297.

rocks the hornblende is seen to have been partially resorbed and changed to augite. The continuation of the resorbtive process until every trace of the hornblende was dissolved, may account for the absence of the mineral in the sheet rocks.

Petrographical Notes.—In an article whose aim is to call forth more accurate determinations of the feldspars in volcanic rocks, and one which gives a practical method for making this determination, Fouqué[8] has described briefly the volcanic rocks of the Upper Auvergne, the acid volcanics of the Isle of Milo and the most important rocks in the Peleponeses and in Santorin. Among the varieties described are doleritic basalts, andesitic basalts, labradorites, andesites, obsidians, trachyte andesites, phonolites. andesitic diabases, rhyolites, dacites and normal basalts. The labradorites are composed largely of microlites of labradorite with a few augites and tiny crystals of olivine in an altered glassy base. In all these cases the author has shown that the rocks contain several different feldspars at the same time, and in each case he has determined their nature. The method made use of in the determination is based on the observation of extinction angles in plates cut perpendicular to the bisectrices.

In a well written article on complementary rocks and radial dykes Pirsson[9] suggests the name of oxyphyre for the acid complementary rock, corresponding to the term lamprophyre for the basic forms. He also calls attention to the fact that the dykes radiating from eruptive centers are usually filled with younger material than that which composes the core at the center. The dykes cutting the central mass will generally be oxyphyres and the more distant ones lamprophyres.

Cordierite gneisses are reported by Katzer[10] from Deutshbrod and Humpolitz in Bohemia, where they are intruded by granite veins, and where masses of them are occasionally completely surrounded by granitic material.

In the examination of a large series of granites and gneisses from the borders of the White Sea, Federow[11] discovered that garnet is present in large quantities when plagioclase is absent and vice versa.

In a general article on the Catoctin belt in Maryland and Virginia, Keith[12] gives very brief descriptions of the granites, quartz porphyries, andesites and the Catoctin schist of the region. The last named rock is apparently a sheared basic volcanic. All the rocks present evidence of having suffered pressure metamorphism.

[8] Bull. Soc. Franc. d. Min., XVII, p. 429.
[9] Amer. Journ. Sci., 1895, p. 116.
[10] Min. u. Petrog. Mitth., XIV, p. 483.
[11] Ibid, p. 550.
[12] 14th Ann. Rep. U. S. Geol. Survey, p. 285.

Reprinted from The American Naturalist, February 1st, 1896.

General Notes.

PETROGRAPHY.[1]

The Eruptives of Missouri.—Haworth[2] has described in much detail the dykes and acid eruptives in the Pilot Knob region, Missouri. The dyke rocks are typical diabases, diabase-porphyrites, quartz-diabase-porphyrites and melaphyres. The author unfortunately classes as

[1] Edited by Dr. W. S. Bayley, Colby University, Waterville, Me.
[2] Mo. Geol. Survey, Vol. VIII, 1895, p. 83-222.

diabase-porphyrites both glassy and holocrystalline rocks. The acid rocks of the region include granites, granite-porphyries, porphyrites and quartz-porphyries. The first two are characteristically granophyric. Their orthoclases are often enlarged by granophyre material whose feldspar is fresh, while the nucleal feldspar is much altered. The quartzes likewise, are enlarged by the addition of quartz around them. There were two periods of crystallization in these rocks. In the second period the phenocrysts were corroded and the groundmass was produced. In addition to the quartz and orthoclase there are present in these rocks also biotite, hornblende, plagioclase and a number of accessory and secondary components. The porphyries and porphyrites contain the same constituents as the granites, from which they are separated simply on account of differences in structure. The phenocrysts are mainly orthoclose, plagioclase, microcline and quartz, many of which are fractured in consequence of magma motions. The groundmass in which these lie is of the usual components of porphyry groundmasses, and in texture is microgranitic, granophyric, micropegmatitic and spherulitic. Many of the porphyries contain fragments of their material surrounded by a matrix of the same composition in which flowage lines are well exhibited. These rocks are evidently volcanic breccias. The author divides the porphyritic rocks into porphyries and porphyrites, the latter containing plagioclase phenocrysts and the former phenocrysts of quartz, orthoclase and microcline.

Rocks from Eastern Africa.—The volcanic rocks of Shoa and the neighborhood of the Gulf of Aden in Eastern Africa comprise a number of varieties that have been carefully studied by Tenne.[2] The main mass of the mountains of the region consists of biotite-muscovite gneiss. This is cut by nepheline basanites, the freshest specimens of which contain phenocrysts of olivine, augite and feldspar in a groundmass of plagioclase, augite, nepheline and often olivine. Trachytes, phonolites and basalts occur in the Peninsula of Aden. The trachytes include fragments of augite-andesite. Inland granophyres with pseudospherulites in their groundmass, trachytes and feldspathic basalts were met with. The granophyres are much altered. In the fine grained product formed by the decomposition of the groundmass of one occurrence quartz, feldspar, and a blue hornblende with the properties of glaucophane can be detected. All the rocks are briefly described. They present no peculiar features other than those indicated.

A Basic Rock derived from Granite.—Associated with the ores in the hematite mines of Jefferson and St. Lawrence Counties, N.

[2] Zeits. d. deutsch. geol. Ges., XLV, p. 451.

Y., is a dark eruptive rock that was called serpentine by Emmons. Smyth[4] (C. H.) has examined it microscopically and has discovered that it consists of a chlorite-like mineral, fragments of quartz and feldspar. By searching carefully he discovered less altered phases of the rock that were identified as granite. The peculiar alteration of an acid granite to a basic chlorite rock is ascribed to chemical agencies. According to the author's notion the pyrite in a neighboring highly pyritiferous gneiss was decomposed, yielding iron sulphates and sulphuric acid. These solutions passed into limestone yielding the ores and then into the granite changing it into chlorite. The altered rock is found only with the ores. The original was probably not always granite. An analysis of the altered rock gave:

SiO_2	Al_2O_3	Fe_2O_3	FeO	MgO	CaO	Na_2O	K_2O	H_2O	Total
29.70	17.03	27.15		10.66	1.68	.56	.10	11.79	= 98.67

Cancrinite-Syenite from Finland.—In the southeastern portion of the Parish Kuolajaroe in Finland, Ramsay and Nyholm[5] secured specimens of a nepheline-syenite containing a large quantity of what the authors regard as original cancrinite. The rock is found associated with gneissoid granite at Pyhakurn. The rock is trachytic in structure and is composed of orthoclase, aegerine, cancrinite and nepheline as essential constituents and apatite, sphene and pyrite as accessories. The cancrinite was the last mineral to crystallize. It occupies the spaces between the other components, and yet it often possesses well defined hexagonal forms. It occurs also as little prisms included within the orthoclase. Because of this association and because the nepheline in the rock is perfectly fresh the cancrinite is regarded as original. This mineral comprises 29.04% of the entire rock.

The same authors in the same paper describe a porphyritic melilite rock found as a loose block a few kilometers W. N.-W. of Lake Wuorijarvi. It contains large porphyritic crystals of melilite, pyroxene and biotite in a groundmass composed of labradorite, zeolites and calcite. The pyroxenes are made up of a colorless augite nucleus surrounded by zones of light green aegerine-augite and deep green aegerine. No olivine was detected in any of the thin sections.

Rocks from the Sweet Grass Hills, Montana.—Weed and Pirsson[6] describe the rocks of the Sweet Grass Hills of Montana as quartz-diorite-porphyrites, quartz-syenite-porphyries and minettes. The first named rock presents no special peculiarities. The quartz-syenite-

[4] Jour. Geology, Vol. 2, p. 667.
[5] Bull. Com. Geol. d. l. Finn., No. 1.
[6] Amer. Jour. Sci., Vol. I, p. 309.

porphyry contains orthoclase, plagioclase and augite-phenocrysts in a fine groundmass of allotriomorphic feldspar and quartz.   The augite is in short thick prisms composed of a pale green diopside core, which passes into a bright green aegerite mantle.   The minette also contains aegerine, but otherwise it is typical.

Petrographical News.—Two peculiar phonolitic rocks are described by Pirsson[7] from near Fort Claggett, Montana. One is a leucite-sodalite-tinguaite, with leucite pseudomorphs, and sodalite as phenocrysts in a groundmass composed mainly of a felt of orthoclase and aegerine.   The leucite pseudomorphs are now an aggregate of orthoclase and nepheline.   In the centers of some of them are small stout prisms of an unknown brown mineral, that is pleochroic in brownish and yellowish tints.   The second rock is a quartz-tinguaite porphyry somewhat similar to Brögger's grorudite.[8]

In a few notes on the surface lava flows associated with the Unkar beds of the Grand Cañon series in the Cañon of the Colorado, Ariz., Iddings[9] briefly describes compact and amygdaloidal basalts and fresh looking dolerites that are identical in all respects with modern rocks of the same character.

Laspeyres[10] estimates that the quantity of carbon-dioxide in liquid and gaseous form contained in rocks is sufficient to serve as the source for all that which escapes from the earth's natural fissures as gas, as well as that which escapes in solution with spring water.   It may be set loose from the rocks through the action of heat or through the action of dynamic forces.

In a handsomely illustrated brochure Merrill[11] describes the characteristics of the onyx marbles and the processes by which they originate. Differences in temperature, according to the author, are not the controlling conditions determining the differences in texture between the onyxes and travertine.   He is inclined to the belief that the banded onyxes were formed by deposition from warm solutions under pressure flowing into pools of quiet cold water.

In a preliminary report on the Geology of Essex County, N. Y. Kemp[12] describes the occurrences of the gneisses, limestones, ophicalcites, gabbros, lamprophyres and other igneous rocks of the district and gives an account of their geological relationships.

[7] Amer. Journ. Sci., 1895, Nov. p. 394.
[8] AMERICAN NATURALIST, 1895, p. 567.
[9] 14th Ann. Rep. U. S. Geol. Survey, p. 520.
[10] Korrespond. bl. Naturh. Ver. preuss. Rheinl., No. 2, 1894, p. 17.
[11] Rep. U. S. Nat. Mus., 1893, p, 539.
[12] Report of State Geologist [of New York] for 1893, p. 433.

Reprinted from The American Naturalist, March 1st, 1896.

General Notes.

PETROGRAPHY.[1]

Examples of Rock Differentiation.—Yogo Peak in the Little Belt Mountains, Montana, consists of a stock of massive igneous rock which breaks up through surrounding horizontal sediments, that have been metamorphosed on their contact with the eruptive. A vertical section through the south face of the mountain caused by a branch of Yogo Creek has afford Weed and Pirsson[2] and excellent opportunity to study the relations of different phases of the eruptive to one another. The massive rock shows a constant variation and gradation in chemical and mineralogical composition along its east and west axis which is two miles in length. In its eastern portion the rock is a syenite, containing pyroxene, hornblende, biotite, orthoclase, oligoclase, quartz and a few accessories. The pyroxene is a pale green diopside and the hornblende a brownish-green variety. The latter is thought to be paramorphic after the former. In structure the syenite is hypidiomorphic with a

[1] Edited by Dr. W. S. Bayley, Colby University, Waterville, Me.
[2] Amer. Journ. Sci., Vol. L, 1895, p. 467.

tendency toward the allotriomorphic structure. Further west, about in the center of the mass, the syenite changes to a darker gray rock with a tinge of green, somewhat resembling a diorite. It is more coarsely crystalline than is the syenite and is much more basic. The minerals are the same as in the syenite, except that quartz is lacking, but differ somewhat in their character and in the proportions present in the two rocks. The augite is now a bright green idiomorphic mineral. Hornblende is rare and biotite abundant. The great difference between this rock, which the authors call yogoite, and the syenite, is in the relative proportions of augite and orthoclase present in them. In the yogoite the pyroxene predominates over the orthoclase, while in the syenite the reverse ratio exists. In the western portion of the rock mass, the prevailing type is shonkinite, a very dark basic rock, very similar to that of Square Butte.[1] Augite and biotite are very abundant as compared with the orthoclase, which in turn predominates over plagioclase. This latter mineral is represented by andesine, a more basic feldspar than that in either the syenite or the yogoite. Analyses of the three types of Yogo Peak rocks follow :

	SiO_2	TiO_2	Al_2O_3	Cr_2O_3	Fe_2O_3	FeO	MnO	MgO	CaO	BaO	SrO	Na_2O	K_2O	H_2O	P_2O_5	Total
Syenite	61.65	.56	15.07	tr	2.03	2.25	.09	3.67	4.61	.27	.10	4.35	4.50	.67	.33	— 100.15
Yogoite	54.42	.80	14.28	tr	3.32	4.13	.10	6 12	7.72	.32	.13	3.44	4 22	.60	.59	— 100.19
Shonkinite	48.98	1.44	12.29	tr	2.88	5.77	.08	9.19	9.65	.43	.08	2.22	4.96	.82	.98	— 99.77

Shonkinite contains in addition .22 per cent. of Fl.

From a consideration of the nature of the three types of rock the authors conclude that the Yogo Peak stock exhibits the results of a progressive differentiation along its major axis. There is a progressive increase in the ferro-magnesian constituents from the east to the west and a consequent increase in basicity. All the components of the three types exhibit the effects of this differentiation in the proportions present in the different rocks. The Yogo Peak mass is thus an illustration of a " Facies suit " as distinguished from a " rock series." In the former differentiation took place in situ, whereas in a ' rock series ' differentiation occurred before the eruption of rocks into their existing positions. The facies suit of Yogo Peak together with the rocks of neighboring mountains comprise a distinct rock series.

The authors close their paper with an appeal for a more specific nomenclature in petrography—a nomenclature that will take account not only of the qualitative relations between the minerals that make up rock masses but of the quantitative relations as well. The Yogo Peak

[1] Compare AMERICAN NATURALIST, 1895, p. 737.

rocks form a natural series with sanidinite and peridotites. Rocks composed of orthoclase and no augite = sanidinite; when orthoclase exceeds augite = augite-syenite; when orthoclase equals augite = yogoite; when augite exceeds orthoclase = shonkinite; when augite alone is present = pyroxenite and peridotite. In this scheme the term augite includes also other ferro-magnesian minerals, and the terms orthoclase other feldspars.

In connection with the article above referred to Iddings[4] mentions the existence of a series of rocks associated with typical basalts and andesites in the Yellowstone National Park. They represent like phases of differentiation belonging to separate, but similar rock families. Most all of these rocks are basaltic looking. They occur in flows and dykes and sometimes as breccias, constituting the major portion of the Absaroka Range. These rocks present a wide range of composition within definite limits, forming a series connected by gradual transitions. Three classes are distinguished, the first of which is characterized usually by abundant phenocrysts of olivine and augite and an absence of feldspar phenocrysts; the second class is characterized by the presence of labradorite phenocrysts in addition to those of olivine and augite, and the third class by the presence of labradorite phenocrysts. The names given to the three classes are absarokite, shoshonite and banakite. The distinctions between the classes is based principally upon their chemical relationships. A large number of analyses, most of which were taken from other papers, illustrate their points of difference. A comparison of the analyses, besides showing the close relationships existing between the rocks of the three classes, shows also what mineralogical differences may obtain for rocks of the same chemical composition. The shoshonite from the base of Bison Peak and the banakite from Ishawooa Canyon have practically the same chemical composition. The former, however, contains abundant phenocrysts of labradorite, augite and olivine, while the latter contains numerous labradorite phenocrysts, but few and small ones of the other two minerals. The groundmass of the first shows much less orthoclase than that of the second, and no biotite, which abounds in the second. The author compares the series of rocks studied by him with the series studied by Merrill[5], with the series discussed by Weed and Pirsson and with Brögger's[6] giorudite-tinguaite series. The conclusion reached by this comparison is to the effect that it may be doubted whether the gen-

[4] Journal of Geology, Vol. III, p. 935

[5] Cf. AMERICAN NATURALIST, 1896, p. 128.

[6] Cf. AMERICAN NATURALIST, 1895, p. 567.

etic relations between igneous rocks can properly mark the lines along which a systematic classification of them may be established.

Petrographical Notes.—In a phyllite-schist found in blocks on the south shore of Lake Michigamme in Michigan, Hobbs[7] has discovered large crystals of a chloritoid like that described by Lane, Keller and Sharpless in 1891. The rock in which the crystals occur is a mass of colorless mica scales through which are distributed large flakes of biotite, small blades of chloritoid, a few acicular crystals of tourmaline and grains of magnetite. Most of the chloritoid is in large porphyritic crystals imbedded in this matrix. The optical properties of the mineral correspond to those of masonite.

In a summary of the results of this work in the upper Odenwald Chelius announces the existence there of two granites—the younger a fine grained aplitic variety and the older a coarse grained porphyritic variety, with a parallel structure due to flowage. Pegmatitic veins that cut this granite are looked upon as linear accumulations of porphyritic feldspar crystals. Many notes are also given on the diorites, gabbros and basalts of the Odenwald, on the basic enclosures in the granite, which the author regards as altered fragments of foreign basic rocks, but nothing of a startling nature with reference to these subjects is recorded. A gabbro porphyry was found occurring as a dyke mass. It consists of phenocrysts of labradorite in a gabbro-aplitic groundmass.

In a general paper on the divisibility of the Laurentian in the Morin area N. W. of Montreal, Canada, Adams[8] describes the characteristics of the members of the Grenville series of gneisses, quartzites and limestones. The augen gneisses, the thinly foliated gneisses and the granulites of the series are all cataclastic or granulitic in structure. They are regarded as squeezed igneous rocks. The crystalline limestones and quartzites are recrystallized rocks that are thought to be changed sedimentaries. Pyroxene gneisses, pyroxene granulites and other allied rocks are of doubtful origin. In addition to all these rocks there is present in the series a group of peculiar banded garnetiferous gneisses which from their chemical composition are regarded as in all probability metamorphosed sedimentary rocks.

[7] Amer. Jour. Sci., Vol. L, 1895, p. 125.
[8] Amer. Jour. Sci., Vol. L, 1895, p. 58.

Reprinted from The American Naturalist, April 1st, 1896.

General Notes.

PETROGRAPHY.[1]

Ancient Volcanics in Michigan.—In an area in Michigan covered by Townships 42 to 47 N. and Ranges 30 to 34 West, is a succession of granites and gneisses overlain by a thickness of some 3000 feet of volcanic rocks, embracing acid and basic flows and tuffs. Among the basic rocks Clements' finds porphyrites and melaphyres, and among the acid ones quartz-porphyries and devitrified rhyolites. The melaphyres and porphyries are described under the names apo-basalts and apo-andesites, because they are altered forms of basalts and andesites. Some of the andesites are amygdaloidal, and nearly all show the effects of pressure. Andesitic and basaltic tuffs are both present. They exhibit no special peculiarities. The quartz porphyries among the acid flows are notable for the existence in them of corroded phenocrysts of quartz in which there has been developed a well marked rhombohedral cleavage. The groundmass of these rocks is sometimes micro-granitic and at other times is micro-poicilitic. The latter structure is peculiar in that it is produced by a reticulating net work of uniformly oriented quartz, between the meshes of which are irregularly shaped areas of orthoclase. The other acid lavas and the acid tuffs are similar to corresponding rocks elsewhere. The series is interesting as affording another illustration of a typical volcanic series of Pre-Cambrian age. It is one of the oldest accumulations of volcanic debris and lavas thus far described.

Gneisses of Essex Co., N. Y.—In a recent bulletin on the geology of Moriah and Westport Townships, Essex Co., N. Y., Kemp' gives a general account of the petrography of the gneisses, limestones, black schists, gabbros, anorthosites and dyke rocks of these regions. Most of these rocks have already been described in more detail in other papers. The gneisses are of several varieties. The most common is a member of the basement complex underlying the other rocks of the district. It is a biotite gneiss composed of quartz, micro-perthite, orthoclase, plagioclase and brown biotite, all of which minerals exhibit evidences of dynamic metamorphism. Near iron ore bodies the gneiss becomes more basic, abundant green or black hornblende, green

---

[1] Edited by Dr. W. S. Bayley, Colby University, Waterville, Me.
[2] Journal of Geology, Vol. III, p. 801.
[3] Bull. N. Y. State Mus., Vol. 3, No. 14, 1895, p. 325.

augite and a large quantity of plagioclase taking the places of the usual gneissic constituents.

Volcanic Rocks in Maine.—In a preliminary notice on the rocks of the Flox Islands, Maine, G. O. Smith[4] gives a brief account of the association of lavas and breccias on North Haven and Vinal Haven Islands. On North Haven the series consists of beds of porphyrites and of coarse volcanic breccias and conglomerates, layers of tuffs and sheets of quartz-porphyry. The porphyrites are sometimes olivinitic. The conglomerates and breccias are composed of fragments of the porphyrites cemented by a porphyritic matrix. The quartz-porphyry possesses no unusual features. On Vinal Haven the rocks are predominantly acid, comprising many banded and spherulitic felsites that were originally glassy rocks. The spherulites are felsitic or fibrous and are certainly original structures, since transitions from the felsitic into brecciated rocks may be traced, in the latter of which occur spherulites that were formed prior to the brecciation. The acid layers of the series are younger than the basic beds.

Spotted Quartzites, S. Dakota.—The Sioux quartzites in Minnehaha Co., S. Dakota, grade upward into variously colored quartz slates that are composed of quartz grains, iron oxides and mica in an argillaceous matrix that has crystallized in part as sericite, kaolin and chlorite. Many of the slates are marked by spots that are lighter than the body of the rocks. These spots are essentially of the same composition as the groundmass in which they lie, except that they contain less iron oxide. Their lighter color is due to bleaching out of the iron salt through the acid, probably of decomposing organic matter.[5]

The Gneisses and 'Leopard Rock' of Ontario.—The gneisses interstratified with the limestones in the Grenville series, north of Ottawa, Canada, vary much in character.[6] The predominant variety is a granitoid aggregate of reddish orthoclase and grayish-white quartz, a little or no mica, and sometimes garnets. Its bedding is very obscure. When the mica is abundant in the rock foliation is distinct. One variety of the rock is called by Gordon syenite-gneiss. It includes the 'leopard rock' of the Canadian geologists. The rock occurs as dykes cutting quartzites and pyroxenites. All the phases of the gneisses show the effects of pressure. The 'leopard rock' consists of

[4] Johns Hopkins Univ. Circulars, No. 121, p. 12.
[5] Beyer: Ib., No. 121, p. 10.
[6] Bull. Geol. Soc. Amer., Vol. 7, p. 95.

ellipsoidal or ovoid masses of feldspar and a little quartz, separated
from each other by narrow anastomosing partitions of green interstitial
substance composed of pyroxene and feldspar.   When the ellipsoids are
flattened by foliation the rock becomes a streaked gneiss.   Under the
microscope, in sections of the coarse grained gneisses, large crystals of
pyroxene, microcline and quartz are seen to be imbedded in a fine
grained aggregate of microcline and quartz.   In the ellipsoidal varieties the ellipsoids are composed mainly of microcline grains and the
interstitial mass is a fine grained mosaic of feldspar, quartz and augite.
In the streaked gneiss the augite is partially changed to green hornblende, while crystalloids of idiomorphic hornblende indicates that some
of this component is an original crystallization.   The rocks are
evidently sheared pyroxene-syenites.   The author discusses the use of
the term 'gneiss' and suggests that the term 'gneissoid' be restricted
to the description of foliated eruptive rocks whose structure is due to
magma motions, that 'gneiss' be used as a suffix to the name of any
rock that has assumed the typical gneissic structure since its original
consolidation, as diorite gneiss, etc., and that the ending 'ic' be used
with reference to the mineralogic composition of a foliated rock whose
origin is unknown—a dioritic gneiss, in this sense indicates a folaited
rock whose present composition is that of a diorite.

Petrographical Notes.—In thin sections of sandstone inclusions
that have been melted by basalts, Rinne[7] finds the remains of quartz
grains surrounded by rims of monoclinic augite, cordierite, spinel, etc.
In some of the glasses formed by the melting of the sandstone are trichites and crystallites of orthorhombic pyroxene.   While this substance
is found abundantly as a contact mineral in the sandstones enclosed in
the basalts of Sababurg, the Blauen Kuppe and Steinberg, the author
nevertheless regards it as a comparatively rare product of the contact
action between these two rocks.

Bauer[8] declares that the rubies, sapphires, spinels and other gem
minerals from northern Burma occur in a metamorphosed limestone on
its contact with an eruptive rock whose nature is not known.

Penfield[9] obtains a heavy solution for the separation of mineral powders whose densities range between 4.6 and 4.94 by melting together
silver and thallium nitrates in different proportions.   The molten mass

[7] Neues Jahrb. f. Min., etc., 1895, II, p. 229.
[8] Sitzb. d. Ges. z. Beförd der gesamnet Naturw.   Marburg, 1896, No. 1.
[9] Amer. Journ. Sci., Dec., 1895, p. 446.

attacks sulphides, but otherwise is of much value in separating mixtures of heavy minerals.

La Touche[10] describes an apparatus to be used in connection with diffusive columns of methylene iodide for the purpose of determining the density of minute fragments of minerals.

[10] Nature, Jan., 1896, p. 198.

Reprinted from The American Naturalist, May 1st, 1896.

General Notes.

PETROGRAPHY.[1]

Malignite, a New Family of Rocks.—Lawson[2] uses the name malignite for a family of basic orthoclase rocks constituting an intrusive mass, possibly laccolitic, in the schists around Poohbah Lake, in the Rainy River district, Ontario. Three phases of the intrusive mass are recognized—a nepheline-pyroxene-malignite, a garnet-pyroxene-malignite and an amphibole-malignite. The constituents common to all phases are orthoclase, aegerine-augite and apatite. In the nepheline variety the nepheline occurs as patches in the orthoclase, or as micropegmatitic intergrowths with it. The orthoclase is in poikilitic relations with all the other minerals, surrounding them like the glass

[1] Edited by Dr. W. S. Bayley, Colby University, Waterville, Me.
[2] Bull. Dept. of Geol. Univ. of California, Vol. I, p, 337.

in a partially crystallized lava.   It was evidently the last component
to solidify.   The composition of the rock is as follows:

SiO$_2$ Al$_2$O$_3$ Fe$_2$O$_3$ FeO CaO MgO Na$_2$O K$_2$O H$_2$O P$_2$O$_5$ Total
47.85 13.24   2.74   2.65 14.36 5.68   3.72   5.25 2.74 2.42 = 100.65

This composition is so similar to that of the Vesuvian leucitophyres,
that the rock is regarded as the plutonic equivalent of these lavas.
The low percentage of silica and the high lime percentage separate the
rock from the eleolite-syenites.

One form of the nepheline-malignite is panidiomorphic through the
development of the orthoclase and all the other components in crystals.
In the garnet-pyroxene phase of the rock the orthoclose is intergrown
with albite in the form of phenocrsyts imbedded in a hypidiomorphic
aggregate of aegerine-augite, melanite, biotite, titanite and apatite.
The augite, melanite and biotite are allotriomorphic.   They seem to
have crystallized contemporaneously with each other, and with a part
of the orthoclase.   In the amphibole-malignite the distinguishing characteristic is the prevalence of a very strongly pleochroic amphibole,
and the absence of any large quantity of aegerine.   The augite that is
present occurs intergrown with the amphibole.   Melanite is wanting,
otherwise this rock is very much like the mellanite-pyroxene malignite.

The author points out the fact that the great mineralogical differences observed in the three types of malignite, are accompanied by
very slight differences in chemical composition.   The three types are
regarded as differentiation phases of the same rock mass.

Foliated Gabbros from the Alps.—Schäfer[1] gives an account
of the olivine gabbro and its dynamically metamorphosed forms which
constitute the rocks of the region in the vicinity of the Allalin glacier
between the Zermatthal and the Saarthal in the Alps.   The normal
gabbro contains in its freshest forms much or little olivine.   In its
altered forms it consists of saussurite, amphibole, talc, actinolite and
garnet.   Ottrelite is often found enclosed in the talc and sometimes imbedded in the saussurite.   In one of the granular varieties of the metamorphosed gabbro a blue amphibole is very abundant.   It is intergrown in part with omphacite.   The granular alteration forms of the
gabbro pass gradually into foliated forms and through these into rocks
called by the author "green schists."   The schistose gabbros are mineralogically similar to the granular alteration phases of the rock,
except that they contain in addition to the minerals named above a

[1] Neues Jahrb. f. Min., etc.. B.B., p. 91.

newly formed albite, zoisite and white mica. The final stage of the alteration is a zoisite-amphibole rock. The green schists are composed of ellipsoids of zoisite, feldspar and epidote imbedded in a schistose green amphibole clinochlor aggregate. Some of the schists are rich in garnets, and others are practically chlorite-schists. All are supposed to be derived from the gabbro.

In addition to the gabbros there are also in the region several exposures of serpentine whose contact with the green schists with which they are associated are always sharp. The original form of the rock is unknown, but it is supposed to have been a peridotite. Its most interesting feature is the possession of light yellow and brown crystals of some member of the humite family.

On the west side of the Matterhorn the author also found normal and olivine gabbros, both more or less altered. The former is cut by little veins of aplite. The peak of the Matterhorn is scarred by numerous fulgurites. Its rocks are fine grained green schists, some of which are like those described above, while others are dense and homogeneous in appearance. They consist of amphibole, clinochlor, zoisite, altered plagioclase, talc and alkali-mica. These rocks are defined as zoisite-amphibolites.

The Rocks of Glacier Bay, Alaska.—Cushing[4] gives a few additional notes on the petrography of the boulders and rocks of Glacier Bay, Alaska. The principal rocks of the region are diorites, altered argillites and limestones that are cut by dykes of igneous rocks. In addition to the diorites and quartz-diorites reported by Williams[5] from this vicinity, there are also in the region mica and actinolite-schists. The dyke rocks are mainly diabases. The author gives some additional information concerning the diorites and briefly describes the schists. The actinolite schists are aggregates of finely fibrous actinolite needles, in whose interpaces is a granular mixture of quartz and epidote and an occasional grain of plagioclase. The mica schists present no unusual features except that some of them are staurolitic.

Petrographical Notes.—As long ago as 1836 Thomson reported the occurrence of light yellowish-green rounded masses which he called huronite, imbedded porphyritically in a boulder of diabase from Drummond Island. Other occurrences of the same substance have been found by the Canadian geologist in diabase dykes cutting the rocks of the Lake Huron region. These have been investigated by Barlow[6]

[4] Trans. N. Y. Acad. Sci., Vol. XV, p. 24.
[5] Cf. AMERICAN NATURALIST, 1892, p. 698.
[6] Ottawa Naturalist, Vol. IX, p. 25.

and are pronounced by him to be aggregates of zoisite, epidote, sericite and chlorite in a mass of basic plagioclase. In other words, huronite is a saussuritized plagioclase. Descriptions of a number of dyke rocks containing 'huronite' are given by this author.

Bauer[7] describes a number of specimens of snow-white, lilac and emerald-green jadeite from Thibet and upper Burmah. One of the green varieties is cut by little veins of nepheline, containing plates of basic plagioclase and little bundles of a monoclinic augite (jadeite) with the same properities as that which constitutes the mass of the jadeite. The rock, according to the author, is made up of this augite and nepheline, the latter mineral acting as a groundmass. The veins are those portions of the rock in which the augite is in very small quantity. In other specimens nepheline occurs in small quantity, and plagioclase is abundant. His conclusion is that the rock is a jadeite-plagioclose-nepheline rock in which locally the one or the other component is most prominent. If the rock is, as the author supposes, a crystalline schist, the occurrence of nepheline in it is of extreme interest.

In a second article the same author[8] describes a serpentine from the jadeite mines at Tauman. It is composed of olivine, picrolite, chrysotite, webskyite and a few other accessories in an albite-hornblende matrix, consisting of an aggregate of single individuals of untwinned albite, in the midst of which lie brown and gray hornblendes surrounded by zones of a bright green variety of the same mineral. Between this zone and the albite there is a fringe of green augite needles. The rocks associated with the jadeite and the serpentine are also described. Among them is a glaucophane-hornblende-schist. All the rocks exhibit the effects of pressure.

In a very short note Beck[9] calls attention to the fact that the molecular volume of dynamically metamorphosed rocks, i. e., of the minerals composing these rocks—is less than that of the original rocks from which they are derived. For instance, a mixture of plagioclase, orthoclase and water in the proportion to form albite, zoisite, muscovite and quartz has a molecular volume of 547.1, while the corresponding mixture of albite, zoisite, etc., has a volume of 462.5,

[7] Neues Jahrb. f. Min., etc., 1896, I, p. 85.
[8] Record of Geol. Survey of India, XXVIII, 3, 1895, p. 91.
[9] Kais. Ak. Wiss. in Wien. Math. Naturw. Class, Jan., 1896.

Reprinted from The American Naturalist, June 1st, 1896.

PETROGRAPHY.[1]

Volcanic Rocks and Tuffs in Prussia.—In the hills east of Ebsdorf, near Marburg, Prussia, are large areas covered by basalt flows, flows of dolerite, and others of rocks intermediate in character between these two, both of which are pre-Tertiary in age, or at any rate are older than the Tertiary beds with which they are associated. The volcanic rocks are cut by dykes of very basic rock resembling limburgite. The little hill west of Wittelsberg, near the northern edge of the basalt area, and the flank of the hill near Kehrenberg, are composed largely of basalt tuff.

The basalt consists of phenocrysts of augite and olivine in a dense felt of augite microlites, biotite and magnetite, in the spaces between which is a colorless glass containing xenomorphic feldspar, leucite and nepheline. Inclusions in the basalt are very common. They comprise besides fragments of foreign rocks, concretions of olivine and of augite. The olivine concretions always contain more or less bronzite, and usually they are surrounded by a violet-brown rim similar to the rims found surrounding the augite phenocrysts in the basalt. Even those concretions that are composed almost exclusively of bronzite are surrounded by rims of this character. The principal component of this rim is a monoclinic augite, so that it appears here that the bronzite, which must have been one of the earliest separations from the magma, was, after its crystallization, changed into augite. Other concretions show the

[1] Edited by Dr. W. S. Bayley, Colby University, Waterville, Me.

alteration of the bronzite into olivine. By complete fusion one concretion, which is thought by the author to have been a bronzite-augite aggregate, has been changed to a mass of rounded augite and olivine grains imbedded in a glass which locally is replaced by nepheline. The alteration of the bronzite, as indicated by the study of a number of sections, is into olivine, augite, magnetite and glass. Among the rare constituents of the olivine concretions are chrome diopside and picotite. The augite concretions or inclusions, consist almost exclusively of a monoclinic augite with which is usually associated a little olivine. In the interiors of the concretions the augite contains fluid enclosures, but toward their peripheries the enclosures are all of glass. Often between the augite grains are little nests of calcite. One of the inclusions observed by the author is abnormal in that it is composed of a small nucleus of augite surrounded by a zone of brown biotite.

Of the foreign inclusions, the author describes two kinds—the calcareous and the granitic. The basalt in the neighborhood of limestone inclusions loses its biotite and magnetite. Nearer the inclusions the augite microlites become light colored and magnetite grains are again developed. At the boundary of the limestone fragment is a rim of large augites, whose ends are directed toward the center of the inclusion. This latter itself is composed of the remnants of calcite grains imbedded in a brown glass, in which are also well formed crystals of a scapolite. The sandstone inclusions have been changed to a mass of quartz grains lying in a brown glass, the whole being surrounded by the usual zone of augite microlites. The granite inclusions first lose their mica. The old feldspar has given rise to newly developed feldspar.

The dolerite seem to occur as a number of small flows that have run together. It presents no special peculiarities. The dyke basalt cutting the tuffs and dolerites sometimes contains well defined crystals of olivine, which occasionally occur as interpenetration twins.

Igneous Rocks of British Columbia.—The petrographical characters of the principal rocks occurring within the area of the Kamloops Map-sheet of British Columbia are described by Ferrier.[2] These rocks embrace feldspathic actinolite schists, diabase porphyrites, harzburgite, amphibolites, diabase tuffs, cherts, gabbros, orthophyres, augite-porphyrites, porphyrites, basalts, pecrite-porphyrites, andesites, trachytes, dacites, diorites, granites, syenites, quartz-porphyries, alnoite and a series of much altered rocks. The descriptions are all brief.

[2] Annual Rep. Geol. Surv. of Canada, Vol. VII, Pt. B., p. 849,

Chalcedony Concretions in Obsidians from Colorado.—
Patton[3] describes the occurrence of large opal and chalcedony concretions or geode-like bodies in beds of a decomposed obsidian on Ute
Creek in Hinsdale Co., Colorado. The concretions are most common
in the upper scoriaceous portions of the flows. Similar concretions
were also found in a rhyolite at Specimen Mountain. The concretions
are composed of radial fibres of chalcedony. The flowage lines that
are common to the rock pass uninterruptedly through them, and in
them are trichites exactly like those in the body of the rock. The concretions are regarded as secondary in origin—and as due to the perco-
lation of silica-bearing waters through the rock. The same author
publishes some photographs of erosion forms produced by the weathering of the volcanic conglomerates in the San Juan Mountains.

Basic Dykes near Lake Memphremagog.—According to
Marsters[4] the Chazy limestones of Lake Memphremagog are cut by
granite, olivine, diabase and lamprophyre dykes. The latter comprise
dark rocks containing phenocrysts of augite, hornblende or olivine.
The olivine, when it occurs, is always situated in the central portions
of the dykes. Sometimes its crystals are one and half inches in diameter. Petrographically these rocks are augite camptonites, fourchites
and monchiquites. The augite camptonite contains both augite and
hornblende in two generations and in varying quantities. Only two
fourchite dykes were observed. Their material presents no unusual
features. The paper is interesting as bringing to our knowledge
another area in which these peculiar and interesting dyke rocks occur.

The Origin of the Maryland Granites.—The last article written by the late Dr. Williams[5] is an introduction to Keyes article on
Maryland granites. In this paper the author explains the criteria by
which ancient plutonic rocks may be recognized in highly metamorphosed terranes, and applies the principles thus established to prove
the eruptive nature of many of the Maryland granites. The pegmatites of the Piedmont plateau were tested by the same criteria, with the
result that these too are pronounced to be eruptive. Many handsome
plates embellish this portion of the paper. In the main portion of the
article Keyes describes the petrographical features of the different types
of granite, giving special attention to the original allanite and epidote
found in them. There is little that is new in the paper, most of its

[3] Proc. Colo. Scient. Soc , Nov. 4, 1895.
[4] Amer. Geol., July, 1895, p. 25.
[5] 15th Ann. Rep. U. S. G. S., 1895, p. 653.

essential points having already been discussed by Hobbs, Grimsley and others.

Petrographical Notes.—The rocks of the Laurentian area to the north and west of St. Jerome, Quebec, are briefly referred to by Adams[6] as gneisses, anorthosites, amphibolites, limestones, quartzites, etc. Some of the gneisses are eruptive and others are probably sedimentary.

Miller and Brock[7] have found in Frontenac, Leeds and Lanark Counties, Ontario, granites, gabbros, scapolite and pyroxene rocks of Laurentian age cut by dykes of quartz gabbro containing phenocrysts of pyroxene and plagioclase.

Keyes[8] declares that the granites and porphyries occuring in the eastern portion of the Ozarks, in Missouri, " are very closely related genetically, and are to be regarded as facies of the same magma," the porphyry being the upper and surface facies of the granite.

[6] Ann. Rep. Geol. Surv. of Can., Vol. VII, J., p. 93.
[7] Can. Record of Science, Oct., 1895.
[8] Bull. Geol. Soc. Amer., Vol. 7, p. 363.

Reprinted from The American Naturalist, July 1st, 1896.

General Notes.

PETROGRAPHY.[1]

The Eruptives and Tuffs of Tetschen.—Two interesting articles on the area of crystalline rocks east of Tetschen on the Elbe, have
appeared simultaneously.   The first, by Hibsch, is a description of
the Tetschen[2] sheet of the map of the Bohemian Mittlegebirges, and the
second by Graber,[3] is on the fragments and bombs occurring in the
tephrite tuffs of the region.

The volcanic rocks of the district are interbedded basalts, tuffites,
tuffs and tephrites, of which the fragmental rocks are in greatest abundance.   Augitites also occur as sheets, and camptonites as dykes in
upper Cretaceous marls.   The older igneous rocks are granitites and
diabases that are associated with clay slates, probably of Cambrian .
age.   Analyses of each of these rocks are given but the rocks are not
described in detail.   The greater portion of the author's article deals
with the volcanic rocks.   The tuffs are composed of basaltic and tephritic fragments of the coarseness of sand in some cases, and in others of

[1] Edited by Dr. W. S. Bayley, Colby University, Waterville, Me.
[2] Min. u. Petrog. Mitth., XV, 1895, p. 201.
[3] Ib., p. 291.

pieces several feet in diameter. These are cemented together by finer portions of the same substances, among which have been deposited zeolites, carbonates, opal and other secondary minerals. Some beds of this tuff are so filled with large fragments of basalt, tephrite, etc., that the rock composing it has been called the "Brocken Tuff." It is to the study of the fragments in this tuff that Graber's paper is devoted.

The basalts and tephrites constitute sheets and lava streams that are interstratified with the tuffs and sediments. Among the former rocks are noticed feldspathic, leucitic and nephelinic varieties, besides in several places magma-basalts. In addition to sheet basalts, dykes and chimneys of this rock have also been observed.

The rocks in all their forms are normal in their development. The author regards contact action around the chimneys as the safest criterion by which to distinguish these forms from denuded sheets and flows. The tephrites comprise hauyn-tephrites, in which hornblende and aegeride are present, nepheline-tephrite, including trachytic and andesitic varieties, and leucite-tephrite composed of phenocrysts of augite, plagioclase and grains of magnetite in a groundmass of these same components, and leucite, biotite and nepheline.

The augite consists of two generations of magnetite and augite in a glassy base. Its analysis gave:

SiO_2	TiO_2	P_2O_5	Al_2O_3	Fe_2O_3	FeO	CaO	MgO	K_2O	Na_2O	H_2O	Moisture	Total
43.35	1.43	1.54	11.46	11.98	2.26	7.76	11.69	.99	3.88	2.41	.59	=99.34

The feldspathic basalt and the andesitic tephrite are the only rocks that seem to have affected the sediments with which they are in contact. Quartzites are changed to aggregates of quartz grains in a glass matrix, where the action is not extremely severe, and to an aggregate of interlocking quartz grains where it has been intense. The article closes with an account of the detailed results of analysis of ten specimens of the volcanic rocks.

Graber's article is devoted principally to a description of the fragments found in the Brocken-tuff. These are all tephritic rocks, among which andesitic, leucitic and phonolitic types are recognized. The characteristics of the components of all these types are portrayed in great detail, especial care being given to the descriptions of the augite and the plagioclase. The phonolitic tephrite is characterized by the presence of nosean, which is in irregular grains. In the andesitic tephrite, which is the most basic variety, the porphyritic augite has an extinction angle $c /_\wedge C$ of 58°–62°, in the leucitic type its extinction is

52°–56° and in the phonolitic type, the most acid variety, it is 50°–53°. In each of the types labradorite and sometimes oligoclase phenocrysts are common, but the feldspar of the groundmass differs in character in the different types. In the andesitic type it is oligoclase, in the leucite variety andesine, and in the phonolitic type sanidine.

A Nepheline-Syenite Bowlder from Ohio.—Miss Bascom[4] has found in the drift near Columbus, Ohio, a bowlder which consists of nepheline-syenite porphyry. The rock is composed of large phenocrysts of oligoclase and smaller ones of nepheline, augite, hornblende and olivine in a groundmass composed of plagioclase and orthoclase laths, hornblende, biotite, augite and magnetite in a feldspathic matrix.

Crystalline Rocks of New Jersey.—In a report on the Archean Highlands of New Jersey, Westgate[5] states that the northern half of Jenny Jump Mt., Warren Co., consists mainly of gneisses with a small area of crystalline limestone, diorites, gneisses, etc. The gneisses are granitoid biotite-hornblende varieties, biotite-gneisset and hornblende-pyroxene gneisses. In the first named variety the prevailing feldspars are microcline and microperthite, and in the pyroxene gneisses plagioclase and orthoclase. The gneisses are cut by pegmatite dykes, amphibolites and diabases.

Associated with the white crystalline limestones are fibrolite and biotite gneisses, hornblendic gneiss, amphibolites, gabbros, norites and diorites, most of the latter of which show evidence of an eruptive origin. Another type of rock often found associated with the limestones is a quartz-pyroxene aggregate, in which the pyroxene is a green or white monoclinic augite. The limestone, the fibrolite and biotite gneisses and the quartz-pyroxene rock are thought to be metamorphosed sediments.

Simple Crystalline Rocks from India and Australia.—Judd[6] gives us an account of several simple crystalline rocks from India and Australia. One is a corundum rock composed principally of corundum grains with rutile, picotite, diaspore and fuchsite as accessory constituents. The corundum is in part pale colored and in part strongly pleochroic. The grains of the latter extinguish together producing with the former a micro-poicilitic structure. One of the specimens examined came from South Rewah and the other from the Mysore State.

[4] Journ. Geol., Vol. IV, p. 160.
[5] Ann. Report State Geol. of New Jersey for 1895. Trenton, New Jersey, 1896, p. 21–61.
[6] Mineralogical Magazine, Vol. XI, p. 56.

Associated with the corundum in the Mysore State is a fibrolite rock. A tourmaline rock from the Kolar gold field in the same State and from North Arcot and Salem in Madras, consists of twisted and bent tourmaline fibres in a matrix of smaller fibres of the same substance. In the neighborhood of Bingera, New South Wales, two rocks are found as dykes cutting serpentine. One consists almost exclusively of green garnets and the other of picotite. The former contains also gold and chrysocolla.

The Weathering of Diabase.—Mr. Merrill[1] describes the changes that have been effected in a grauular diabase at Medford, Mass., during its disintegration into soil. Bulk analysis of the fresh and the weathered rock yielded the following results:

	SiO_2	Al_2O_3	Fe_2O_3	FeO	CaO	MgO	MnO	K_2O	Na_2O	P_2O_5	Ign	Total
Fresh	47.28	20.22	3.66	8.89	7.09	3.17	.77	2.16	3.94	.68	2.73	=100.59
Weathered	44.44	23.19	12.70		6.03	2.82	.52	1.75	3.93	.70	3.73	= 99.81

The disintegration of the rock is accompanied by a leaching out of its most soluble constituents. Assuming that the alumnia has remained unchanged in quantity in the course of the disintegration, the percentage of each constituent lost in this process is shown to be as follows:

SiO_2	Al_2O_3	Fe_2O_3	FeO	CaO	MgO	MnO	K_2O	Na_2O	P_2O_5	Ign
18.03	.00	18.10		25.89	21.70	41.57	29.15	12.83	11.39	.09

The paper is full of valuable suggestions that cannot be even referred to in these notes.

Petrographical Notes.—Transitions from massive anorthosites into augen gneisses and into thinly foliated gneisses and transitions from olivine gabbro into hornblende schists are briefly described by Kemp[8] in a preliminary article on the dynamic metamorphism of anorthosites and related rocks in the Adirondacks.

Pirsson[9] suggests the use of the word anhedron to express the meaning usually expressed in the phrase ' hypidiomorphic form.' An anhedron is a body with the physical constitution and properties of a crystal but without the crystallographic form. The term may be conveniently applied to the crystalline grains in rock masses.

[1] Bull. Geol. Soc. Amer., Vol. 7, p. 349.
[8] Bull. Geol. Soc. Amer., Vol. 7, p. 488.
[9] Ib., Vol., p. 492.

Reprinted from The American Naturalist, August 1st, 1896.

PETROGRAPHY.[1]

Petrography of the Bearpaw Mountains, Montana.—The Bearpaw Mountains are the dissected remains of a group of Tertiary volcanoes. Their cores of the old volcanoes are granular rocks, their lavas and tuffs are represented by basic sheets and beds. The lavas are largely basalts, leucite-basalt and other similar basic types.[2]

The cores consist of mica-trachytes, quartz-syenite, porphyries, containing aegerite-augite and anothoclase-phenocrysts, in which are imbedded microlites of oligoclase, trachytes containing hornblende and diopside and shonkinite. A few miles from Bearpaw Peak a denuded core is exposed, which furnishes a good example of the differentiation of a syenite in place. The intrusion is laccolitic in character. Around its borders it has highly altered the sedimentary rocks with which it is in contact. The most acid portion of the laccolite is a light aplitic syenite containing quartz and diopside. The main mass is a more basic syenite resembling monzonite or yogoite. It contains diopside and much plagioclase. The most basic phase is a shonkinite. Analyses for the three principal types follow :

	SiO_2	Al_2O_3	Fe_2O_3	FeO	MgO	CaO	Na_2O	K_2O	H_2O_5	Other	Total
Quartz-syenite	68.34	15.32	1.90	.84	.54	.92	5.45	5.62	.45	.57 =	99.95
Monzonite	52.81	15.66	3.06	4.76	4.99	7.57	3.60	4.84	1.09	1.86 =	100.24
Shonkinite	50.00	9.87	3.46	5.01	11.92	8.31	2.41	5.02	1.33	2.68 =	100.01

The totals corrected for Fe and Ce are 99.94, 100.22 and 99.93 respectively.

Two French Rocks.—In the serpentine of St. Préjet-Armadon, Haute-Loire, France, Lacrou[3] finds nodules composed of asbestiform gedrite surrounding a kernel of serpentine or biotite. The nodules are separated from the serpentine by an envelope of biotite. They are sup-

---

[1] Edited by Dr. W. S. Bayley, Colby University, Waterville, Me.
[2] Weed and Pirson : Amer. Journ. Sci., IV, Vol. 1, p. 283 and 351.
[3] Bull. Soc. Franc. d. Min., XIX, p. 687.

posed to be of secondary origin.   Bronzite and asbestus both occur in
the rock.   In the norite area of Arvien, Auvergne, the same author
describes a variety of this rock which is characterized by the presence
of secondary reaction, rims of anthophyllite and actinolite between its
hypersthene and plagioclase, the former appearing next to the pyroxene.   The plagioclase of the rock is often altered to actinolite, garnet
and albite, while the hypersthene is changed to an aggregate of anthophyllite.

The Granite of the Himalayas.—McMahon[4] describes the
granite of the N. A. Himalayas.   Although highly foliated in the borders of its masses, the rock is shown to be eruptive.   The author thinks
the foliation is due to pressure upon the rock before it finally solidified.
He attempted to show that this schistosity could not possibly have been
produced after the rock cooled.   The granite is coarsely porphyritic
with large orthoclase crystals in a medium to fine grained groundmass
composed of the usual constituents of granite.   This is cut by tiny
veins of quartz which are supposed to represent the micrystallized residue left after the first partial consolidation of the rock, or to be the
result of a partial fusion of the quartz grains originally occurring in it.
This quartz, though it presents the usual aspects of secondary quartz,
is thought to have been injected into the vein spaces while it was in a
molten condition.   Sinuous areas and viens of microcrystalline mica
are likewise observed in the granite, and these are thought to have been
produced by the rapid crystallization of mica that had been melted,
and not by the crushing and shearing of the original micas nor by secondary processes of any other kind.   The paper is well illustrated by
photo-micrographs.

California Rocks.—Fairbanks[5] describes the rocks of Eastern California between Mono Lake and the Mojave desert as comprising both
sedimentary and igneous forms.   Among the latter are both granitic
and volcanic varieties.   The granites form the eastern slope of the
Sierra Nevadas.   In the northern portion of the area it is a coarsely
porphyritic biotite hornblende variety.   In the southern portion it is
replaced by a more basic phase containing less hornblende.   The volcanic rocks met with in the district are andesitic flows, dykes and tuffs,
and basalt flows among the more recent rocks and liparites among the
more ancient ones.   The microscopical description of the type is
deferred to a later paper.

[4] Proc. Geologists Association, Vol. XIV, p. 287.
[5] Amer. Geologist, Vol XVII, p.'63.

Turner[6] gives a classification of the igneous rocks studied by himself from various places in California. He divides them into families in accordance with their mineralogical composition, including in the same family all those rocks with the same composition irrespective of structure. He then takes up the syenites and discusses them in some detail. The family is made to include syenites (granular), syenite-por-phyries (porphyritic) and trachytes (microlitic and glassy) and apo-trachytes. The syenites include soda-syenite or albitite, augite-syenite, hornblende-syenite and mica-syenite. The apo-trachytes include among other rocks Rosenbusch's orthophyres and keratophyres. Until very recently no rocks of the syenite family have been proven to occur within the borders of the State. All those rocks described as such are now known to be hornblende-andesites, granites or diorites. The author refers briefly to the known occurrence of the syenites in the State and describes more fully some new ones.

He reports dykes of white albitite-porphyries or soda-syenite porphy-ries in the rocks of the Mother lode quartz mines. In the bed of Moc-casin Creek the rock consists of quartz, muscovite and albite, but in other places it consists almost exclusively of albite with a few grains of an olivine-green mineral thought to be aegerite. The rock resem-bles somewhat Brögger's sölosbergite and Palache's albite rock contain-ing crossite. An analysis of one specimen gave:

SiO_2	TiO_2	Al_2O_3	Fe_2O_3	FeO	CaO	MgO	K_2O	Na_2O	H_2O	P_2O_5	Total
67.53	.07	18.57	1.13	.08	.55	.24	.10	11.50	.46	.11	=100.34

Gabbro-Gneiss from Russell.—The gabbro of Russell, St. Lawrence Co., N. Y., is said by Smyth[7] to change its character rapidly in consequence of a variation in grain from moderately fine to very coarse, in structure from porphyritic to granular and in color from black to gray. Upon alteration the gabbro passes into a rock made up of red masses in a groundmass of gabbro. In other places it becomes schistose, when it takes on a granulitic texture. Sometimes hornblende is developed in it in long narrow plates that run approximately at right angles to the schistosity, causing the rock to resemble a metamor-phosed sediment. Even in the most gabbroitic varieties of the rock the plagioclase is changed into an aggregate of secondary products, among which scapolite is the most common. In the change of the massive gabbro into the schistose variety the constituents are first

[6] Ib., Vol. XVII, p. 375.
[7] Amer. Jour. Sci., Vol. 1, p. 273.

grauulitized and then drawn out into lenticular areas. The feldspars of the gneisses appear to have been recrystallized, since the feldspathic areas consist of single feldspar individuals and not fragments of grains. The pyroxene also differs from the gabbro pyroxene. It has lost its characteristic black inclusions and has assumed a deep green color. This mineral, as well as the hornblende, which is abundant in the gneisses, are both regarded as having recrystallized, the augite material coming from the original augite of the gabbro and the hornblende from the secondary amphibole so common in the gabbro. The gneisses are thus schistose gabbros in which recrystallization has taken place with attendant granulitization. The author points out the fact that in the first stages in the alteration of the gabbro scaly hornblende and scapo-lite are formed, while in the final stage they have completely disap-peared, and in this latter stage there results a gneiss which bears no evidence of having been crushed.

Notes.—The serpentine near Bryn Mawr, Penna., has resulted by the alteration of a peridotite according to Miss Bascom.[8] The rock of the Conshohocken dyke is a typical diabase.

[8] Proc. Amer. Acad. Science, 1890, p. 220.

Reprinted from The American Naturalist, September 1st, 1896.

PETROGRAPHY.[1]

Geology of Point Sal, California.—The geology of Point Sal, the extreme northwestern corner of Santa Barbara County, California, has been carefully worked out by Fairbanks[2] with special reference to the igneous rocks found there. The sedimentary rocks constituting the point and the adjacent country are of miocene or later age. They

[1] Edited by Dr. W. S. Bayley, Colby University, Waterville, Me.
[2] Bull. Dep. Geol. Univ. of Cal., Vol. 2, p. 1.

comprise volcanic ashes, gypsiferous clays and bituminous shales, the last named of which were regarded by Lawson as tuffs. The present author declares them to be organic deposits. The igneous rocks which penetrate these beds are all basic. They include gabbros, peridotites, basalts, diabases and rocks similar to those heretofore described as analcite diabases. These latter are all now considered by the author as representing the otherwise practically unknown type of the teschenites. The augitic variety of this rock has the general structure of the diabases, in which are large poikilitic plates of augite. Between the diabasic constituents are polyhedral grains of analcite, and, in what appear to have been cavities in the rock-mass, are little groups of crystals and crystalline masses of the same mineral. The plagioclase in the rock is all zonal with nuclei of labrodorite surrounded by concentric zones of a more and more acid feldspar, the peripheral one being albite. An analysis of a coarse grained specimen gave:

SiO_2 Al_2O_3 Fe_2O_3 FeO CaO MgO K_2O Na_2O P_2O_5 Ign.   Total
49.61 19.18   2.12   5.01 10.05 4.94 1.04   5.62   .27   3.55=101.39

which corresponds very nearly to 43.3 per cent feldspar, with a density of 2.57, 32.3 per cent augite, 20 per cent analcite, 4 per cent magnetite and .04 per cent apatite. All of the analcite is supposed to be an alteration product of nepheline.

The basalts of the region include two types. One is the usual variety and the other an amygdaloidal and spheroidal variety that is intruded by diabases and diabasic gabbros. These last named rocks grade into one another. Both contain hornblende, some of which is regarded as secondary and some as primary. In addition to the diabasic-gabbros there are others associated with peridotites (and serpentines) in such a manner that both rocks are regarded as differentiated products of the same magma. The gabbro is sometimes massive. At other times it is possessed of a gneissic structure, often attended by a striping produced by the alternation of augitic and feldspathic bands. The structure is concluded, after study, to be the result of stretching.

Among the other basic rocks identified in the gabbro-peridotite complex are anorthosites, diorites, norites, lherzolites, picrites, saxonites, wehrlites, dunites and pyroxenites. Each type is well described and a discussion of the banding noticed in many of them is given in some detail.

Leucite-Basanites of Vulcanello.—After studying carefully the rocks on Vulcanello in the Lispari Islands, Bäckström[3] concludes

[3] Geol. För. i Stockh. Förhanl., XVIII, p. 155.

that the greater portion of them are leucite-basanites. They all contain phenocrysts of augite, labradorite, olivine and magnetite in a groundmass which is sometimes a holocrystalline aggregate of oligoclase, orthoclase, leucite and magnetite, and at other times of numerous leucites, small augites and iron oxides in a glassy matrix. The rocks are regarded as effusive types of lamprophyres (minettes or kersantites) a supposition which is the more probable from the fact that the effusives in the Lipari province are mainly feldspathic basalts, andesites, liparites and trachytes. Biotite and leucite are thought to be complementary minerals—the former separating from a siliceous magma under considerable pressure, and the latter from a magma of the same composition under surface pressure, under conditions favorable to the escape of the mineralizers fluorine and water. Leucite is not confined to rocks rich in potash, nor is it necessarily characteristic of these. Its place may often be taken by biotite.

A Squeezed Quartz-Porphyry.—A squeezed quartz-porphyry is described by Sederholm[4] as occurring at two places in the Parish of Karvia in Province Abo, Finland. In both it appears as dykes cutting granite. The rock consists mainly of microcline phenocrysts to which are often added growths of new microcline in optical continuity with the original crystals, phenocrysts of an acid plagioclase surrounded in many cases by microcline substance and quartz phenocrysts in a groundmass of orthoclase and quartz. The twinning of the microcline is more largely developed around quartz enclosures in the phenocrysts and near quartz veins than elsewhere in the crystals. The porphyritic quartzes occasionally retain their dehenhedral contours, but usually they are much deformed in outline and in their optical characteristics. Often the quartzes are so shattered that they now constitute lenticular areas of a quartz mosaic. The structure of the groundmass is in several types. In the most important one it consists of a micropegmatite of orthoclase and quartz containing shreds of chlorite, which in some cases are distributed so as to exhibit a fluidal arrangement. The granite through which the porphyry cuts is a coarse grained porphyritic variety composed of oligoclase, biotite and hornblende. On the contact with the dyke rocks it is crushed and much epidote is developed in it. Under the microscope it presents the usual aspects of a dynamically metamorphosed rock. In his discussion concerning the name to be applied to the porphyry, the author quotes from a letter by Dr. Williams in which the prefix ' apo ' is defined as signifying that the rock

[4] Bull. Com. Geol. d'Finlande, No. 2, 1895.

to which it refers has been changed from its original character through devitrification.

Mica-Syenites at Rothschonberg.—Two dykes of mica-syenite cut the phyllite formation near Rothschönberg, Saxony, producing in the neighboring rocks contact metamorphism. One of the dykes weathers spheroidally, and in the kernels of the spheroids fresh material for study was afforded Henderson,[5] who found the rock to be composed of orthoclase, plagioclase, quartz, biotite, apatite and several accessory components. The feldspar and quartz both occur in grains and in crystals, the biotite in flakes. An analysis of the rock gave the figures below (I).

The second occurrence differs little from the first. Muscovite is present as well as biotite, otherwise the two rocks are practically alike in mineral composition. Its chemical composition is shown in (II).

	SiO_2	Al_2O_3	Fe_2O_3	CaO	MgO	K_2O	Na_2O	H_2O	CO_2	S	Total
I.	61.40	16.66	7.46	2.08	3.65	2.93	4.75	.76	1.54	.20	=101.43
II.	57.63	16.47	5.37	5.25	4.44	3.12	5.15	.45	2.14	.95	=100.97

The structure of both rocks was panidiomorphic, although the development of secondary quartz renders them now hypidiomorphic. They are syenitic aplites. In the neighboring phyllites new biotite has been abundantly developed and hornblende has been produced in some quantity. The free silica which is abundant in the unaltered phyllites has become combined with metallic elements in the altered forms. While the percentage of silica in specimens taken at 2 meters and 11 meters from the contact and at the contact is the same, the free quartz in the first is 43.38 per cent of the rock's mass, in the second 38.94 per cent and in the third 34.06 per cent.

Reprinted from The American Naturalist, October 1st, 1896.

General Notes.

PETROGRAPHY.[1]

The Sioux Quartzite of Iowa.—The Sioux quartzite has long been known as the oldest sedimentary rock in Iowa. It has recently been studied by Beyer.[2] It is a white or red vitreous rock with which is associated as its upper extension a series of mottled reddish or purplish-black slates. The quartzites present the usual aspects of indurated sandstones. The constituent quartz grains are rich in 'quartz-needles' which can be traced directly into rutile spicules. The slates are arenaceous. They exhibit no traces of slaty cleavage, though in some cases their quartz grains and micaceous constituents are distorted in such a way as to testify to a horizontal movement in the rock mass containing them. All the slates are mottled by spheroidal masses of a lighter color than the body of the rock. These masses are spheroidal with the longer dimensions of the spheroids in the bedding planes of the shale. Their lighter color is supposed to be due to the removal of iron from those portions of the rock they occupy. Associated with the quartzites is a great mass of olivine diabase consisting of a coarse grained aggregate of labradorite and oligoclase zonally intergrown, olivine, augite, biotite, hornblende, apatite and magnetite. Most specimens are much altered, the components having been changed into the usual secondary substances common to diabase. In structure the rock varies from the ophitic, in which the plagioclase is older than the augite, to the gabbroitic, in which the augite is the older mineral. An analysis gave:

SiO_2	TiO_2	Fe_2O_3	FeO	Al_2O_3	CaO	MgO	K_2O	Na_2O	H_2O	P_2O_5	Total
42·85	tr	13.66		20.23	6.85	3.42	1.90	5.78	.88	tr	=100.57

The Peridotites of North Carolina.—In connection with a discussion of the occurrence and origin of corundum in North Carolina, Lewis[3] gives us an interesting account of the basic rocks associated with the gneisses in that portion of the Appalachian belt included within the limits of the State. These basic rocks, consisting mainly of

[1] Edited by Dr. W. S. Bayley, Colby University, Waterville, Me.
[2] Iowa Geol. Survey, Vol. VI, p. 69.
[3] Bull. No. 11, North Carolina Geol. Survey, 1896.

peridotites, occur in small lenticular masses or in narrow strips, which
are always enveloped in a sheet of schistose talc or chlorite, and thus
are never in direct contact with the gneisses through which they are
believed to cut.   They are classed as peridotites, pyroxenites and amphibolites, the former being the most common.   The peridotites present several types in each occurrence, all merging into one another and
forming a single geological unit.   The principal types of the peridotites are dunite, harzburgite, amphibole-picrite and forellenstein.   All
are massive, as a rule, though exceptions are noted.   The dunite is composed of olivine grains, octahedrons and rounded grains of picotite and
chromite, plates of enstatite, prisms of light green hornblende and
various alteration products of these, the most common being serpentine tremolite and chlorite.   The Harzburgite and the other peridotites present no unusual features.   They appear to be transition phases
between the dunite and the various pyroxenites among which are recognized two types, an enstatite rock and websterite.   The enstatite
rock is made up almost exclusively of enstatite or bronzite and its alteration product talc.   An analysis of the enstatite gave :

SiO_2	Al_2O_3	FeO	CaO	MgO	MnO	H_2O	Total
51.64	.12	9.28	.45	31.93	.56	5.45	99.43

The amphibolites are composed chiefly of amphibole.   The most important type is composed of grass-green hornblende, anorthite and
more or less corundum.   The rock is fine grained and it is usually
gneissic, although occasionally massive.   Transitions through forellenstein into dunite were observed, although the distribution of the rock
suggests its occurrence in a system of dykes cutting the latter rock.
The hornblende has the following composition :

SiO_2	Al_2O_3	Cr_2O_3	FeO	NiO	MgO	CaO	NaO	K_2O	H_2O	Total
45.14	17.59	.79	3.45	.21	16.69	12.51	2.25	.36	1.34 =	100.33

Genth called the mineral smaragdite.   Dana regards it as edenite.   In
addition to the rocks mentioned above, there are also present in the
region massive serpenite, which was unquestionably derived from dunite, talc-schists, and soapstones derived from enstatite rocks and chlorite
schists.

In a second paper[4] the same author gives his reasons for considering
these rocks as eruptive in origin.

Shales and Slates from Wales.—Hutchins[5] continues his
studies of clays, shales and slates by an investigation of the nature of

[4] Elisha Mitchell Sci. Soc. Jour., Pt. II, 1895, p. 24.
[5] Geol. Mag., Vol. III, 1896, p.

shales taken from some of the deepest coal mines in Wales. The chemical composition of the particular shale analyzed does not differ materially from that of some of the carboniferous shales from other coal fields. Physically the deeper shales are not much more compact than hard clays. The author reviews the results of his observations on shales and slates. He states that what takes place in a rock during its progress from clay to shale, is the development and crystallization of muscovitic mica and the production of chlorite. He also calls attention to the fact that dynamic metamorphism is made to explain many phenomena connected with the crystallization of slates, that are capable of being explained better by static metamorphism. The spots of many contact rocks are now thought to be secretions from a mineralizing solution, depositing in these spherical forms material collected from the rock body. By crystallization the spots pass over into cordierite, biotite or staurolite crystals.

Notes.—Cushing[6] declares that in addition to the rocks described by Kemp from the eastern Adirondacks there is a system of diabase dykes, which are older than the monchiquites and camptonites of the district.

By melting certain rock powders in the presence of reagents Schmutz[7] has obtained aggregates of minerals which in most cases are very different from those composing the original rocks. Eklogite fused in the presence of calcium and sodium fluride yielded a mass of meionite, plagioclase and glass; leucitite with calcium chloride gave a mass composed of a glassy groundmass and plagioclase; with the addition of sodium fluride and potassium silico-fluride it yielded scapolite, mica, magnetite; with sodium chloride it produced augite, scapolite and magnetite and a glass matrix. Granite fused with magnesium and calcium chlorides and sodium fluoride gave andesine and olivine in a groundmass containing augite. Other rocks treated with other reagents gave analagous results.

As the result of a series of experiments made with the view of discovering a medium with a very high specific gravity that will not attack sulphides, Retgers[8] finds that the acetate and the mixed nitrate and acetate of thallium are both neutral toward sulphides. The former is available for separating minerals with a density below 3.9, and the latter those with a density below 4.5.

[6] Trans. N. Y. Acad. Sci., XV, 1896, p. 249.
[7] Neues Jahrb. f. Min., etc., 1896, I, p. 211.
[8] Neues Jahrb. f. Min., etc., 1896, I, p. 213.

In an article in the *Neues Jahrbuch* Bauer[9] gives a German transcription of his article[10] on the rocks associated with the jadeite of Turmaw, Burmah.

Schroeder vander Kolk[11] describes briefly a series of rocks collected by Martin in the Moluccas. In the southern part of Amboina the rocks are mainly granite and peridotite, while in the larger northern part they consist of modern volcanics, as they do also on the other islands studied. These rocks are principally dacites and liparites, but on one island andesites occur. Both the dacites and the granite contain cordierite. The dacites are pyroxene and biotite varieties. The andesites are pyroxenic; mica schists, breccias and limestones occur also on the islands. The residue left after treatment of the limestone with acid contains quartz, sanidine, plagioclase, biotite, amphibole, orthorhombic pyroxene, hematite, garnet, cordierite, sillimanite and pleonost.

[9] Neues Jahrb. f. Min., etc., 1866, I, p. 19.
[10] Cf. AMERICAN NATURALIST, June, 1896, p. 478.
[11] Ib., 1896, I, p. 152.

Reprinted from The American Naturalist, December 1st, 1896.

PARTIAL INDEX OF SUBJECTS.

INDEX OF AUTHORS.

Lightning Source UK Ltd.
Milton Keynes UK
UKHW012249140219
337323UK00011B/670/P